Interactions Between Exercise Physiology and Metabolism

Interactions Between Exercise Physiology and Metabolism

Guest Editors

Lijing Gong
Enming Zhang

MDPI

Basel • Beijing • Wuhan • Barcelona • Belgrade • Novi Sad • Cluj • Manchester

Guest Editors
Lijing Gong
China Institute of Sport and Health Science
Beijing Sport University
Beijing
China

Enming Zhang
Department of Clinical Sciences Malmö
Lund University
Malmö
Sweden

Editorial Office
MDPI AG
Grosspeteranlage 5
4052 Basel, Switzerland

This is a reprint of the Special Issue, published open access by the journal *Metabolites* (ISSN 2218-1989), freely accessible at: https://www.mdpi.com/journal/metabolites/special_issues/3P95RJ6477.

For citation purposes, cite each article independently as indicated on the article page online and as indicated below:

Lastname, A.A.; Lastname, B.B. Article Title. *Journal Name* **Year**, *Volume Number*, Page Range.

ISBN 978-3-7258-7731-7 (Hbk)
ISBN 978-3-7258-7732-4 (PDF)
https://doi.org/10.3390/books978-3-7258-7732-4

Contents

About the Editors

Lijing Gong

Lijing Gong is an associate researcher at the China Institute of Sport and Health Science at Beijing Sport University and at the Key Laboratory of Exercise and Physical Fitness under the Ministry of Education of China. Her recent research has primarily focused on sports and health promotion, including the mechanisms by which exercise improves chronic diseases such as diabetes, reduces the toxicity of cancer chemotherapy drugs, and the mechanisms underlying altitude/hypoxia training and rehabilitation after hypoxic injury. Her other research interests encompass sports nutrition and the effects of nutritional supplements on enhancing exercise performance and alleviating exercise fatigue.

Enming Zhang

Enming Zhang is an Associate Professor at the Department of Clinical Sciences at the Malmö, Faculty of Medicine, Lund University, Sweden, and a Principal Investigator at NanoLund: Centre for Nanoscience, Lund University. Over the past two decades, his research has focused on elucidating the roles of cellular molecular sensors in pancreatic beta cells and on understanding how their activity can be precisely regulated. His work integrates multidisciplinary approaches, including molecular biology, genetics, and nanotechnology. His expertise spans islet biology, magnetic biology, calcium signaling, and advanced cellular imaging.

Preface

This Reprint, titled Interactions between Exercise Physiology and Metabolism, brings together one editorial and eight original research articles published in *Metabolites*. It represents a comprehensive effort to advance our understanding of how physical activity modulates systemic metabolism at the molecular, cellular, and whole-organism levels.

The subject of this Reprint lies at the intersection of exercise physiology and metabolic regulation. Its scope spans from large-scale clinical cohort analyses and exercise calorimetry studies to mechanistic investigations using multi-omic approaches, including metabolomics, microbiomics, epigenomics, and sportomics. The research covers diverse populations—from elite athletes and collegiate sprinters to elderly urban residents and animal models of neurodevelopmental disorders—highlighting the broad translational potential of exercise science.

The aim of this Reprint is to provide a systems-level perspective on how exercise influences host metabolism, gut microbiome composition, epigenetic regulation, and inter-organ communication. A central purpose is to move beyond traditional physiological markers and embrace cutting-edge analytical platforms that capture the dynamic complexity of exercise-induced metabolic adaptations. By integrating findings from studies on hypertension management, metabolic syndrome, athletic performance, autism spectrum disorder, and lifestyle intervention epigenetics, this collection seeks to establish a scientific foundation for precision exercise medicine.

The motivation for compiling this work stems from the urgent global need to counteract the epidemic of chronic metabolic diseases driven by sedentary lifestyles and nutritional imbalances. As a non-pharmacological, cost-effective, and accessible intervention, exercise holds immense promise for preventing and managing conditions such as obesity, type 2 diabetes, cardiovascular disease, and even neurodevelopmental disorders. However, realizing this promise requires a deep mechanistic understanding of how different exercise modalities, intensities, and durations affect metabolic pathways in diverse individuals. The research presented here addresses this gap by employing state-of-the-art technologies to decode the molecular signatures of exercise.

This Reprint is addressed to a broad audience, including exercise physiologists, metabolomics researchers, sports scientists, clinical nutritionists, endocrinologists, and public health professionals. It will also serve as a valuable resource for graduate students and postdoctoral fellows seeking to apply multi-omics approaches to exercise and metabolism research. We hope that this collection inspires further investigation into the molecular mechanisms of exercise and facilitates the translation of these insights into personalized, evidence-based exercise prescriptions for improving human health.

Lijing Gong and Enming Zhang
Guest Editors

metabolites

MDPI

Editorial

Interactions Between Exercise Physiology and Metabolism

Lijing Gong [1,2,3], Chunyan Xu [4] and Enming Zhang [5,*]

1 Key Laboratory of Exercise and Physical Fitness, Ministry of Education, Beijing Sport University, Beijing 100084, China; lijing.gong@bsu.edu.cn
2 Key Laboratory for Performance Training & Recovery of General Administration of Sport, Beijing Sport University, Beijing 100084, China
3 China Institute of Sport and Health Science, Beijing Sport University, Beijing 100084, China
4 Beijing Higher School Engineering Research Center of Sport Nutrition, Beijing Sport University, Beijing 100084, China; xucy@bsu.edu.cn
5 Department of Clinical Sciences in Malmo, Lund University Diabetes Centre, Lund University, 20213 Malmo, Sweden
* Correspondence: enming.zhang@med.lu.se

Physical inactivity and excess nutritional intake have contributed to a global pandemic of metabolic and non-communicable diseases, resulting in unhealthy fat accumulation, metabolic dysregulation, and a high burden of chronic conditions such as type 2 diabetes, cardiovascular disease, and obesity [1]. Exercise, as a potent, non-pharmacological, and cost-effective intervention, orchestrates a complex network of physiological adaptations. It stimulates the release of a plethora of signaling molecules, including exerkines, which participate in inter-organ communication, regulate numerous key signaling pathways, and, in turn, enhance whole physical health, help prevent the development of chronic diseases, and actively help the body to recover [2]. The field of exercise metabolism seeks to capture these complex processes, and with the support of advanced omics technologies, it has revolutionized our ability to map the molecular landscape of exercise adaptations [3].

This Special Issue, entitled "Interactions between Exercise Physiology and Metabolism", presents a compelling collection of eight original research articles that explore the multi-faceted relationships between physical activity and metabolic health. The contributions cover a wide range of fields, ranging from large-scale epidemiological analyses in clinical populations to mechanistic studies in animal models and elite athletes. By utilizing state-of-the-art metabolomics, microbiomics, and epigenomics, these studies provide novel insights into how exercise modulates host metabolism, gut microbiota, and epigenes, as well as how these interactions influence health promotion and disease treatment.

New Paradigms in Exercise Metabolism Research: From Traditional Indicators to Systems Biology

Several studies in this Special Issue highlight the powerful potential of systems biology approaches in exercise metabolism research. Zhao et al. conducted a large-scale cohort study of 3373 urban elderly residents, employing statistical methods such as piecewise structural equation modeling (PiecewiseSEM) to systematically elucidate the mediating role of serum lipids in the relationship between physical activity (PA) and blood pressure (BP). Their findings revealed that the mediating effect of serum lipids was particularly significant under low-PA conditions. This discovery addresses the limitations of traditional studies that only focus on the direct association between PA and BP, offering a novel theoretical perspective for hypertension management (Contribution 1).

On the other hand, Fei et al. utilized untargeted plasma metabolomics to find that cardiorespiratory fitness (CRF) levels can mitigate the metabolic disturbances associated

 https://doi.org/10.3390/metabo16040229

with metabolic syndrome (MetS) risk factors by modulating key metabolic pathways such as the TCA cycle and arginine biosynthesis. The eight common differential metabolites identified in this study (including branched-chain amino acids and 2-oxoglutarate) provide potential biomarkers for stratifying CRF levels and MetS severity (Contribution 2).

Collectively, these studies illustrate how macroscopic epidemiological analysis or microscopic metabolite profiling can yield new insight into the complex relationships between exercise and metabolism.

Exercise Interventions and Metabolic Disease Management: From Mechanisms to Applications

The role of exercise as a non-pharmacological intervention in metabolic disease management is strongly emphasized in this Special Issue. Varlet et al. conducted a large-scale calorimetry study of 6465 individuals, providing the first systematic characterization of the linear relationship between carbohydrate oxidation and exercise intensity, termed the "carbohydrate cost of the watt (CCW)." Their findings revealed that women, older individuals, and those with higher adiposity exhibited higher CCW values, suggesting a greater reliance on carbohydrates and relatively lower lipid oxidation capacity. This discovery provides a physiological basis for individualized exercise prescription in patients with obesity and metabolic diseases (Contribution 3).

Bathina et al. further investigated the mechanisms of lifestyle intervention (diet combined with exercise) from an epigenetic perspective. In severely obese men with hypogonadism, a 12-month lifestyle intervention significantly reduced global DNA methylation levels in peripheral blood mononuclear cells and downregulated the expression of key DNA methyltransferases (DNMT1, DNMT3A, DNMT3B) as well as adipogenic genes (PPARγ, CEBPα, FTO). This study provides novel evidence that lifestyle interventions can improve metabolic status through epigenetic reprogramming, offering a molecular explanation for the long-term benefits of exercise (Contribution 4).

Training Modalities and Athletic Performance: Metabolic Adaptations and Individual Difference

Studies involving athletic populations provide unique insights into the metabolic adaptation mechanisms of different training modalities. Kong et al. compared the effects of a threshold training model (72%:24%:4% intensity distribution) and a polarized training model (78%:8%:14%) on adolescent male rowers. Although both training modalities significantly improved 2 km rowing performance, the metabolic profiles they induced differed: the threshold group specifically enriched the pyruvate metabolism pathway, while the polarized group enriched the aminoacyl-tRNA biosynthesis pathway. This finding suggests that while both modalities are equally effective in enhancing aerobic endurance, they modulate through different molecular mechanisms, providing a metabolic rationale for scientific training (Contribution 5).

Fu et al. adopted a "sportomics" strategy to systematically monitor the multi-dimensional physiological status of collegiate sprinters during the pre-competition strengthening and tapering periods. Despite reduced training volume during tapering, athletes exhibited gut microbiome imbalance, elevated inflammatory factors (IL-6, IL-1β), and activation of immune-related metabolic pathways. These subtle physiological changes, not fully captured by traditional biochemical markers, were closely associated with athletic performance limitations, underscoring the unique value of multi-omics monitoring in elite sports (Contribution 6).

Li et al. further revealed the impact of sex differences on exercise-induced metabolic responses. Through serum metabolic profiling of freestyle wrestlers at different training stages, they observed that male athletes mainly exhibited increases in branched-chain amino acids and lactate during peak training, whereas female athletes showed decreased

https://doi.org/10.3390/metabo16040229

valine and increased pyruvate levels. This finding suggest that sex-specific metabolic responses should be fully considered in training monitoring (Contribution 7).

The Exercise–Gut–Multi-Organ Axis: From Peripheral Metabolism to the Central Nervous System

The gut microbiome has recently emerged as a critical mediator connecting exercise and systemic metabolism. The studies included in this Special Issue extend this perspective to the field of neural development. Meng et al. utilized a Shank3 knockout rat model of autism spectrum disorder (ASD) and provided the first evidence that early-life swimming intervention can reverse striatal metabolic abnormalities. Shank3-knockout resulted in significant reductions in neurotransmitters and neurodevelopment-related metabolites such as glutamate, glutamine, and taurine, whereas early swimming intervention restored these metabolite levels and improved glutamatergic and GABAergic synaptic pathways. This study offers a mechanics insight into the beneficial effects of exercise on ASD symptoms and supports the regulatory role of the exercise–gut–brain axis (Contribution 8).

Consistent with this perspective, Fu et al. also demonstrated that during the tapering period, athletes exhibited a decreased gut microbiome health index (GMHI), an increased microbial dysbiosis index (MDI), and alterations in inflammation- and immune-related metabolic pathways. These findings further support the gut microbiome as a sensitive indicator of physiological responses to exercise (Contribution 6).

Conclusions and Future Perspectives

The studies presented in this Special Issue collectively provide a multifaceted and comprehensive view of the interactions between exercise and metabolism. Encompassing large-scale population cohorts, clinical metabolic diseases, elite athletes' performance, and neurodevelopmental disorders, and integrating approaches from traditional biochemical assessments to advanced multi-omics analyses, these contributions significantly deepen our understanding of exercise metabolism mechanisms.

Future research should prioritize the integration of multi-dimensional data, including metabolomics, metagenomics, and epigenomics, combined with advanced analytical techniques such as artificial intelligence, to develop predictive models for individualized exercise prescriptions. Such efforts will facilitate the transformation of exercise interventions from a "one-size-fits-all" approach toward precision and personalization. Moreover, enhancing the translation of basic research findings into clinical practice and validating the utility of exercise-related metabolic biomarkers in disease prevention and management will be essential for addressing the growing global challenge of metabolic diseases.

Author Contributions: L.G. wrote the original draft of manuscript; C.X. and E.Z. reviewed and edited the manuscript. All authors have read and agreed to the published version of the manuscript.

Funding: This research was funded by the China Fundamental Research Funds for the Central Universities (Beijing Sport University File No. 2024TZJK001, 2025KYPT05).

Institutional Review Board Statement: Not applicable.

Informed Consent Statement: Not applicable.

Data Availability Statement: No new data were created or analyzed in this study. Data sharing is not applicable to this article.

Conflicts of Interest: The authors declare no conflicts of interest.

https://doi.org/10.3390/metabo16040229

List of Contributions

1. Zhao, Y.; Zhang, Y.; Wang, F. Mediating Effects of Serum Lipids and Physical Activity on Hypertension Management of Urban Elderly Residents in China. *Metabolites* **2024**, *14*, 707. https://doi.org/10.3390/metabo14120707.
2. Fei, X.; Huang, Q.; Lin, J. Plasma Metabolomics Study on the Impact of Different CRF Levels on MetS Risk Factors. *Metabolites* **2024**, *14*, 415. https://doi.org/10.3390/metabo14080415.
3. Brun, J.-F.; Varlet, E.; Myzia, J.; Varlet-Marie, E.; Raynaud de Mauverger, E.; Mercier, J. Carbohydrate and Fat Oxidation in Muscle Assessed with Exercise Calorimetry in 6465 Subjects. *Metabolites* **2026**, *16*, 121. https://doi.org/10.3390/metabo16020121.
4. Bathina, S.; Fuenmayor Lopez, V.; Prado, M.; Teo, S.B.; Villareal, D.T.; Chen, R.; Qualls, C.; Armamento-Villareal, R. Lifestyle Intervention Therapy Modulates Global DNA Methylation and Adipogenic Gene Expression in Severely Obese Hypogonadal Men. *Metabolites* **2026**, *16*, 198. https://doi.org/10.3390/metabo16030198.
5. Kong, F.; Zhu, M.; Pan, X.; Zhao, L.; Yang, S.; Zhuo, J.; Peng, C.; Li, D.; Mi, J. The Metabolome Characteristics of Aerobic Endurance Development in Adolescent Male Rowers Using Polarized and Threshold Model: An Original Research. *Metabolites* **2025**, *15*, 17. https://doi.org/10.3390/metabo15010017.
6. Fu, P.; Duan, X.; Zhang, Y.; Dou, X.; Gong, L. Based on Sportomics: Comparison of Physiological Status of Collegiate Sprinters in Different Pre-Competition Preparation Periods. *Metabolites* **2024**, *14*, 527. https://doi.org/10.3390/metabo14100527.
7. Li, X.; Liu, X.; Liu, J.; Liu, Y.; Han, Y.; Zhang, W. Serum Metabolomics Reveals Metabolic Changes in Freestyle Wrestlers During Different Training Stages. *Metabolites* **2025**, *15*, 737. https://doi.org/10.3390/metabo15110737.
8. Meng, Y.; Hu, Y.; Xue, Y.; Zhen, Z. Metabolomic Profiling of the Striatum in Shank3 Knockout ASD Rats: Effects of Early Swimming Regulation. *Metabolites* **2025**, *15*, 134. https://doi.org/10.3390/metabo15020134.

References

1. Khoramipour, K.; Sandbakk, O.; Keshteli, A.H.; Gaeini, A.A.; Wishart, D.S.; Chamari, K. Metabolomics in Exercise and Sports: A Systematic Review. *Sports Med.* **2022**, *52*, 547–583. [CrossRef]
2. Chow, L.S.; Gerszten, R.E.; Taylor, J.M.; Pedersen, B.K.; van Praag, H.; Trappe, S.; Febbraio, M.A.; Galassetti, P.R.; Gao, Y.; Haus, J.M.; et al. Exerkines in Health, Resilience and Disease. *Nat. Rev. Endocrinol.* **2022**, *18*, 273–289. [CrossRef]
3. MoTrPAC Study Group; Lead Analysts; MoTrPAC Study Group. Temporal Dynamics of the Multi-Omic Response to Endurance Exercise Training. *Nature* **2024**, *629*, 174–183. [CrossRef] [PubMed]

metabolites

MDPI

Article

Mediating Effects of Serum Lipids and Physical Activity on Hypertension Management of Urban Elderly Residents in China

Yang Zhao [1,2], Yike Zhang [1] and Fei Wang [1,2,*]

[1] Sports Science Institute, Shanxi University, Taiyuan 030006, China; zhaoyang@sxu.edu.cn (Yang Zhao)
[2] School of Physical Education, Shanxi University, Taiyuan 030006, China
* Correspondence: nemo@sxu.edu.cn

Abstract: Background/Objectives: Investigating the importance and potential causal effects of serum lipid biomarkers in the management of hypertension is vital, as these factors positively impact the prevention and control of cardiovascular disease (CVD). **Methods**: We surveyed 3373 urban residents using longitudinal data from the CHARLS database, collected between 2015 and 2020. Pearson correlation methods were employed to explore the relationships among the numerical variables. A logistic regression model was utilized to identify the risk factors for hypertension. The dose–effect relationship between serum lipids and BP was assessed using restricted cubic splines (RCS). Additionally, piecewise structural equation modeling (PiecewiseSEM) was conducted to further elucidate the direct and indirect pathways involving individual body indices, serum lipids, and PA on BP responses at different levels of physical activity (PA). **Results**: The four serum lipids showed significant differences between hypertensive and non-hypertensive residents ($p < 0.05$). All lipids, except for HDL cholesterol, demonstrated extremely significant positive correlations with both systolic blood pressure (SBP) and diastolic blood pressure (DBP) ($p < 0.001$). All serum lipid variables were significantly associated with the incidence of hypertension. Specifically, triglycerides (bl_tg), HDL (bl_hdl), and low-density lipoprotein LDL cholesterol were identified as significant risk factors, with odds ratios (ORs) of 1.56 (95% CI: 1.33–1.85, $p < 0.001$), 1.16 (95% CI: 1.02–1.33, $p < 0.05$), and 1.62 (95% CI: 1.23–2.15, $p < 0.001$), respectively. Conversely, cholesterol (bl_cho) was a protective factor for hypertension, with an OR of 0.60 (95% CI: 0.42–0.82, $p < 0.01$). PA showed weak relationships with blood pressure (BP); however, PA levels had significant effects, particularly at low PA levels. The four serum lipids had the most mediating effect on BP, especially under low PA level conditions, while PA exhibited a partly weak mediating effect on BP, particularly under high PA level conditions. **Conclusions**: Serum lipids have significant nonlinear relationships with BP and PA levels exert different influences on BP. The significant mediating effects of serum lipids and the weak mediating effects of PA on individual body indices related to SBP and DBP demonstrate significant differences across varying levels of PA, highlighting the importance of low PA levels in hypertension management. This study could provide valuable recommendations and guidance in these areas.

Keywords: serum lipids; physical activity; SEM; hypertension management; urban elderly residents; China

1. Introduction

As society has evolved, the lifestyles of urban residents have shifted significantly, leading to a high prevalence of chronic diseases [1]. Hypertension, defined as a systolic blood pressure (SBP) of 140 mmHg or higher and/or a diastolic blood pressure (DBP) of 90 mmHg or higher, is a prevalent condition that affects a significant portion of the population and is a major risk factor for cardiovascular disease [2]. The prevalence of hypertension in China has increased from 5.1% to 23.2% over the past three decades [3,4], and the number rose to 56.8% in northeast China in the 2020s [5]. The number of hypertensive patients aged 30 to 79 years has doubled from 1990 to 2019 [6] and is projected to reach

1.56 billion by 2025 [7]. Meanwhile, the number of fatalities caused by hypertension among Chinese residents was 2.54 million in 2017, with cardiovascular disease deaths accounting for 95.7% [8]. Hypertension has become a significant public health concern worldwide. Studies have shown that it is the leading risk factor contributing to the global burden of disease [9]. In China, hypertension sharply increases with age and is a significant risk factor for ischemic heart disease, stroke, chronic kidney disease, and dementia [4]. It imposes a heavy burden on families and society.

Due to common risk factors or disease pathogenesis, hypertension often coexists with other chronic diseases, such as dyslipidemia [10]. The results of a 2018 national survey indicated that the prevalence of dyslipidemia among adults over 18 years old was 35.6% [11]. Hypertension and dyslipidemia are the most common risk factors for cardiovascular disease (CVD) [12]. The coexistence of hypertension and dyslipidemia significantly increases the harmful effects on the cardiovascular system [13]. The result of this interaction is an exponential increase in CVD. Moreover, controlling BP and serum lipids has become a priority for the prevention and management of CVD. Hypertension, dyslipidemia, and obesity are independent risk factors for CVD, and they often co-exist [12]. These conditions represent a serious medical issue that significantly increases the risk of heart, brain, and other diseases. In China, 23.2% of adults aged 18 and above have hypertension [3]. Dyslipidemia is recognized as an independent risk factor for atherosclerosis, coronary heart disease, and stroke. During three national cross-sectional surveys conducted from 2002 to 2015, serum lipid levels increased sequentially among Chinese adults [4].

Physical activity (PA) has proven to be an effective tool for the primary and secondary prevention of several chronic diseases [14]. The intensity of PA also plays a role, as moderate- to high-intensity activities have been shown to significantly reduce BP in hypertensive patients and lower the risk of CVD [15–17]. A study of 1311 African Americans found that exercise-related PA had the potential to lower the risk of developing high BP, while other types of activity were not associated with this risk [18]. Another investigation involving 125,402 adults found that longer commuting times and high-intensity leisure activities were significantly associated with a lower risk of hypertension, especially among older individuals [19]. Most previous studies have examined the effects of PA on BP. However, current global estimates indicate that one in four adults and 81% of adolescents do not engage in the recommended levels of PA as suggested by the WHO [20]. In China, PA was observed to be in sharp decline while urbanization was increasing dramatically [21]. Additionally, some studies have examined the impact of age, gender, waist circumference, and BMI on the development of hypertension, revealing a positive correlation between these factors and BP [22]. The study by Wang et al. [23] revealed the significant impact of body shape control on managing hypertension. All these variables were incorporated into the statistical modeling to explore their impact on BP. While PA has beneficial effects on the chronic diseases mentioned above, the mechanism by which PA influences BP through serum lipid metabolites remains unclear. Furthermore, the sequence of effects between PA and serum lipids in managing hypertension has not been reported. Investigating the significance of serum lipid biomarkers in the context of PA for hypertension management is essential for lifestyle interventions aimed at preventing cardiovascular disease (CVD) and improving overall health. Additionally, understanding the pathways through which PA and blood lipids influence CVD risk factors can enhance prevention and disease control strategies. Mediation analysis is commonly used to explore the mechanisms of interventions, and epidemiologists have increasingly employed it to analyze mechanisms focused on path analysis in recent years. However, studies examining the mechanisms by which PA and serum lipids affect BP are limited, particularly concerning the internal pathway effects among different constructs. The objective of the study is (1) to clarify the relationship between BP and PA, serum lipids, and other relevant variables; (2) to quantify the effects of PA and serum lipids on BP; and (3) to assess the underlying mechanisms among PA, serum lipids, and BP. This study aims to provide theoretical support for research on hypertension in elderly urban residents and offer valuable insights into hypertension management.

2. Materials and Methods

In this study, we selected residents from the China Health and Retirement Longitudinal Study (CHARLS) databases from 2015 to 2020. It is a longitudinal database that began in 2008, with individuals followed up every two years. All data will be made public one year after the completion of data collection. In our study, the updated database extends to 2020, and recently, new data delayed by COVID-19 have been published. The information collected by CHARLS includes highly scientific and standardized health and medical measurements and provides valuable data on social and economic environments. These data have demonstrated significant contributions to research. Detailed information regarding the study design and sampling strategy can be found in the cohort information [24]. In the data extraction process, we first extracted variables including height, weight, age, gender, smoking and drinking habits, urban/rural residency, waist circumference, sleep duration, PA, and four serum lipids. We then filtered the data to include only individuals with systolic blood pressure (SBP) $\geq$ 140 mmHg, and/or diastolic blood pressure (DBP) $\geq$ 90 mmHg, and/or those using blood pressure medication, as well as individuals aged over 60. Finally, we removed any missing values for these variables, resulting in a final sample of 3373 elderly residents aged 60 years and older. All participants provided informed consent, and CHARLS was approved by the Institutional Review Board of Peking University (Code: IRB00001052-11015).

The serum lipid data included triglycerides (bl-tg), high-density lipoprotein cholesterol (bl-hdl), low-density lipoprotein cholesterol (bl-ldl), and total cholesterol (bl-tc). The final sample comprised 1649 individuals with hypertension and 1724 without hypertension. Among these respondents, there were 2663 females and 710 males.

Metabolic equivalent (MET) was used to calculate PA intensity.

$$\text{PA} = \sum_i (\text{MET}_i \times \text{F}_i \times \text{T}_i) \tag{1}$$

where F_i is the weekly frequency of activity of the PA level (day/week); T_i is the activity time of day on the PA level (minutes/day); MET_i is the constant related to the PA level. The PA levels are divided into three types: vigorous PA (VPA), moderate PA (MPA), and light PA (LPA). According to the Chinese Guidelines of the International Physical Activity Questionnaire (IPAQ) [25], the MET for VPA, MPA, and LPA were assigned 8.0, 4.0, and 3.3, respectively. The PA duration is categorized as follows: "1 = (<0.5 h), 2 = (0.5 h ~ 2 h), 3 = (2 h ~ 4 h), 4 = ($\geq$4 h)" [26]. In this study, PA intensity was ultimately categorized into three levels according to the recommendation from the World Health Organization (WHO) [27,28]: Low ($\leq$10 MET·h/week), Moderate (10 ~ 50 MET·h/week), and High ($\geq$50 MET·h/week) levels.

The demographic data of participants are given in Table 1. Continuous numeric variables are presented as the mean $\pm$ standard deviation, while categorical variables are presented as frequency and percentage. Spearman correlation analysis was used to determine the relationship between BP and all other numeric variables. A logistic regression model was employed to identify protective and risk factors for hypertension. The dose–response effects of serum lipids and BP were evaluated using restricted cubic splines (RCS) for non-linear correlations and piecewise linear regression to identify inflection points, considering significance (p < 0.05). We used the piecewiseSEM R package for structural equation modeling (SEM) to further assess the associations among serum lipids (bl_tg, bl_hdl, bl_ldl, and bl_cho), PA, and individual body indices (height, weight, and waist) affecting SBP and DBP, respectively, after accounting for multiple key factors such as age and sleep time. In SEM, we used path analysis to explore potential causal relationships between variables. These relationships are represented by line segments, where an arrow pointing to a variable indicates that another variable influences it, and the number on the line represents the path coefficient. All measured variables included in this model were first standardized and then included in the SEM. To confirm the robustness of the relationships between serum lipids and PA, we used piecewiseSEM to account for random effects related to hypertension, providing the "marginal" and "conditional" contributions of serum lipids and PA predictors in driving model stability.

Table 1. Statistical tables of the baseline characteristics of the population.

Characteristic [n/N(%) | Mean (SD)] [1]	Unit	Hypertension?		p-Value [2]
		not, N = 1724	yes, N = 1649	
Gender				0.606
Male	—	369/1724 (21%)	341/1649 (21%)	
Female	—	1355/1724 (79%)	1308/1649 (79%)	
PA_level				0.015
Low	—	1248/1724 (72%)	1157/1649 (70%)	
Moderate	—	274/1724 (16%)	322/1649 (20%)	
High	—	202/1724 (12%)	170/1649 (10%)	
Smoke				0.764
Yes	—	126/1724 (7.3%)	125/1649 (7.6%)	
Not	—	1598/1724 (93%)	1524/1649 (92%)	
Drink	—			0.206
Yes	—	382/1724 (22%)	336/1649 (20%)	
Not	—	1342/1724 (78%)	1313/1649 (80%)	
Urban				<0.001
Yes	—	648/1724 (38%)	724/1649 (44%)	
Not	—	1076/1724 (62%)	925/1649 (56%)	
Age	years	66.74 (5.93)	68.42 (6.40)	<0.001
Height	cm	154.00 (22.27)	153.81 (22.93)	0.465
Weight	kg	55.24 (25.01)	59.43 (12.22)	<0.001
Waist	cm	83.30 (25.85)	88.54 (13.51)	<0.001
PA	MET·h/w	15.62 (29.92)	16.20 (28.95)	0.223
Sleep_time	h	6.19 (2.09)	6.08 (2.14)	0.116
Systole	mmHg	119.35 (11.92)	146.42 (18.85)	<0.001
Diastole	mmHg	69.27 (8.09)	80.29 (11.21)	<0.001
bl_tg	µg/mL	131.88 (80.28)	155.38 (90.09)	<0.001
bl_hdl	µg/mL	53.11 (11.37)	51.17 (11.10)	<0.001
bl_ldl	µg/mL	105.63 (27.86)	107.78 (29.44)	0.023
bl_cho	µg/mL	188.72 (36.11)	192.14 (36.38)	0.004

Notes: [1] Mean (SD) for numeric variables; n/N (%) for categorical variables; [2] Pearson's Chi-squared test for categorical variables; Kruskal–Wallis rank sum test for numeric variables.

All statistical analyses and diagrams were conducted using R (version 4.2.1, https://cran.r-project.org, accessed on 1 July 2022). In the data cleaning process, the timestamp was handled using the "janitor" and "lubridate" packages in R. The "tidymodels" package was used for data cleaning and preprocessing of all predictors. Descriptive analysis was performed using the "tbl_summary" package, while Pearson's correlation analysis utilized the "corr" package. Logistic regression was carried out using the "log" package. The RCS analysis was conducted using the "ggrcs" package. The SEM analyses were performed using the "piecewiseSEM," "nlme," and "lme4" packages. We used Fisher's C test (when $0.05 < p < 1.00$) to confirm the goodness of fit of the modeling results. We then modified our models based on significance ($p < 0.05$) and the overall goodness of fit. All plots were generated using the "ggplot2" package.

3. Results

3.1. Descriptive Statistics

Residents with hypertension exhibited significant differences from non-hypertensive residents in individual categorical variables and most continuous variables (Table 1). Among the continuous variables, four serum lipid levels showed significant differences between hypertensive and non-hypertensive residents ($p < 0.05$). Except for bl-hdl, higher values were observed in hypertensive residents. PA showed relatively higher in hypertensive residents, though differences in total PA absolute values were not statistically significant compared to non-hypertensive residents. However, significant differences were found across different PA levels ($p < 0.05$). In terms of continuous variables, the waist circumference of hypertensive patients was measured at 88.54 ± 13.51 cm, significantly higher than that of non-hypertensive patients, which was 83.30 ± 25.85 cm ($p < 0.001$). Additionally, other variables such as weight (59.43 ± 12.22 kg for hypertensive patients compared to 55.24 ± 25.01 kg for non-hypertensive patients) and age (68.42 ± 6.40 years for hypertensive patients compared to 66.74 ± 5.93 years for non-hypertensive patients) were also higher in hypertensive patients. Both height and sleep duration showed no significant differences between the two groups ($p > 0.05$). Furthermore, among the categorical variables, only the residency status (urban vs. rural) exhibited significant differences ($p < 0.001$) between hypertensive and non-hypertensive residents.

3.2. Pearson Correlation Analysis

The Pearson correlation analysis was used to determine the relationships among PA, four serum lipids variables, SBP, DBP, and various numerical variables (Figure 1). As shown in Figure 1, age, waist circumference, weight, and four serum lipids—except for high-density lipoprotein cholesterol (bl-hdl)—demonstrated extremely significant positive correlations with both SBP and DBP ($p < 0.001$). The serum lipid of bl-hdl showed a significant negative correlation with SBP ($p < 0.05$). Additionally, bl-hdl, triglycerides (bl-tg), and age exhibited more significant correlations with other numerical variables. Sleep time only showed significant correlations with two serum lipids: bl-hdl and total cholesterol (bl-cho).

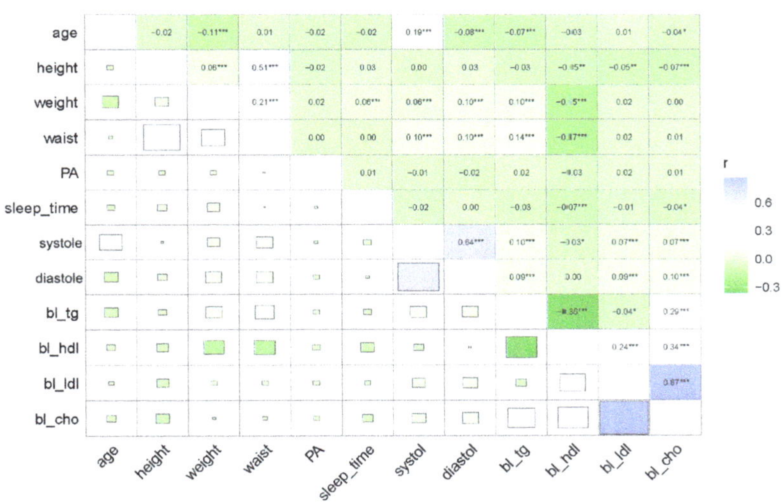

Figure 1. Spearman correlation analysis of four serum lipids variables, PA, SBP, DBP, and various numerical variables: The numbers represent correlation coefficients, and markers represent significance levels as * $p < 0.05$, ** $p < 0.01$, and *** $p < 0.001$.

3.3. Logistic Regression Analysis

Residents, both with and without hypertension, were included as a binary dependent variable in the study. Binary logistic regression analysis was conducted to identify potential risk factors for hypertension. Figure 2 illustrates the significant influencing factors on hypertension based on the logistic regression model.

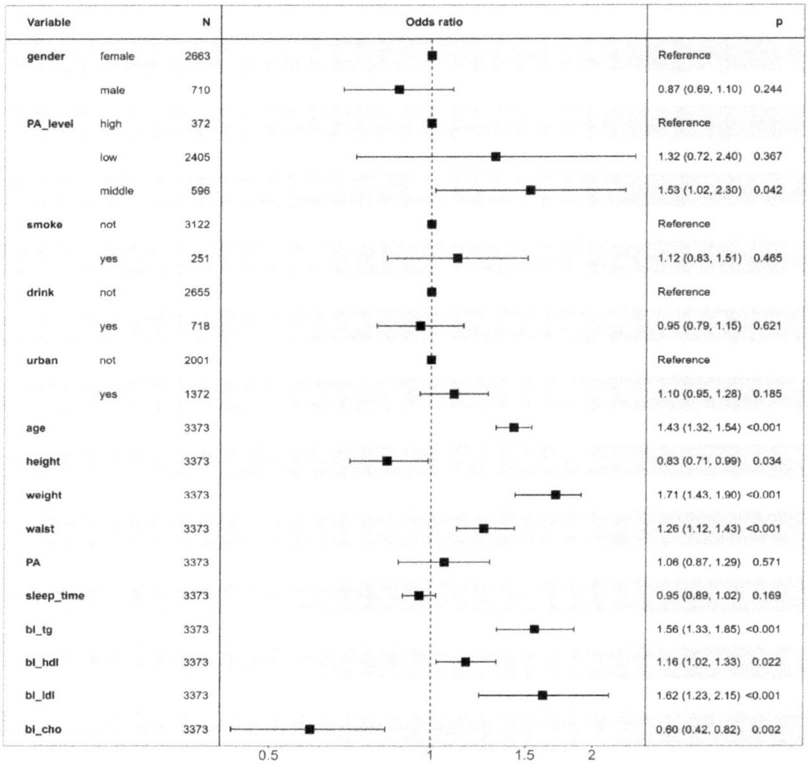

Figure 2. Factors affecting hypertension based on logical regression model.

The results indicated that three serum lipids—bl_ldl (OR = 1.62, 95% CI: 1.23–2.15, $p < 0.001$), bl_tg (OR = 1.56, 95% CI: 1.33–1.85, $p < 0.001$), and bl_hdl (OR = 1.16, 95% CI: 1.02–1.33, $p < 0.05$)—were potential risk factors for hypertension. Additionally, one serum lipid, bl_cho (OR = 0.60, 95% CI: 0.42–0.82, $p < 0.01$), was identified as a potential protective factor. Moreover, numeric variables such as age, waist circumference, and weight were also significant risk factors for hypertension.

In terms of categorical variables, having a moderate PA level compared to a high PA level was identified as a risk factor for hypertension (OR = 1.53, 95% CI: 1.02–2.30, $p < 0.05$), while other categorical variables were not identified by the logistic regression model. The result indicates that, compared to residents with high PA levels, those with moderate PA levels faced a 53% increased risk of developing hypertension. Conversely, the numerical variable of height was found to be a slight protective factor against hypertension (OR = 0.83, 95% CI: 0.71–0.99, $p < 0.05$).

3.4. Dose–Effects of Serum Lipids with BP

Our analysis initially indicated a potential non-linear relationship between serum lipids and both SBP and DBP. Utilizing RCS analysis, we identified a non-linear relationship between serum lipids and BP (Figure 3a for SBP and Figure 3b for DBP). The breakpoint that defines the natural spline was 4. In the RCS model, we adjusted the covariables including age, waist circumference, height, weight, sleep time, and PA. Age showed a highly significant positive effect on both SBP and DBP ($p < 0.001$). However, sleep time showed a significant negative effect on SBP ($p < 0.05$). The baseline body indices of height, weight, and waist circumference also demonstrated highly significant positive effects on both SBP and DBP ($p < 0.001$). In addition, the effects of serum lipids showed similar trends for both SBP and DBP, with the highest effects observed for baseline high-density lipoprotein (bl_hdl), followed by baseline low-density lipoprotein (bl_ldl), baseline cholesterol (bl_cho), and baseline triglycerides (bl_tg). This suggests that the relationship between serum lipids and both SBP and DBP is not simply linear but exhibits an inverted U-shaped relationship.

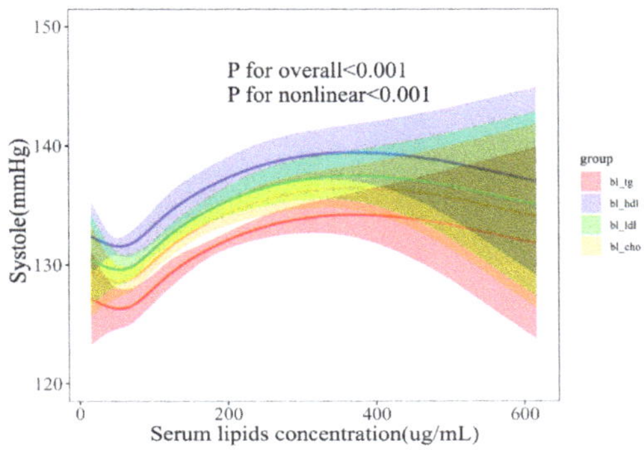

(**a**)

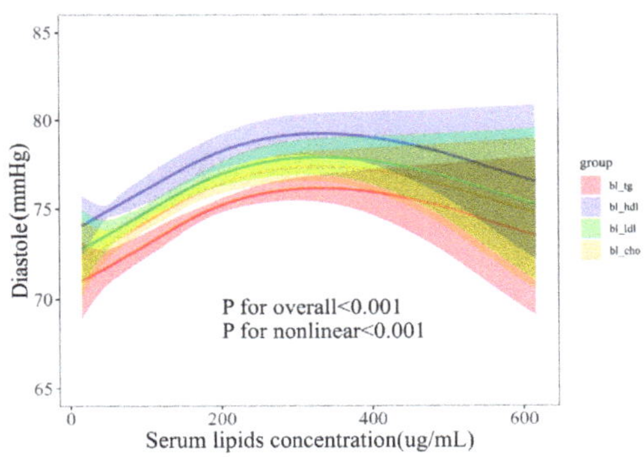

(**b**)

Figure 3. Dose–effects of serum lipids on systole (**a**) and diastole (**b**).

3.5. Mediating Effects of PA and Serum Lipids on BP

Piecewise SEM was performed to uncover the direct and indirect pathways through which individual baseline characteristics, serum lipids, PA, age, and sleep duration have effects on blood pressure (BP) responses at different PA levels. The direct and indirect pathways by which these regulatory factors impact variations in SBP and DBP are illustrated in Figure 4. The combined influence of individual baseline characteristics, serum lipids, PA, age, and sleep duration accounted for a significant proportion of SBP variations at low PA levels (58%), moderate PA levels (57%), and high PA levels (65%), considering the random effects of "Hypertension?" (blue sections of Figure 4). Similarly, these variables explained a substantial proportion of DBP variations at low PA levels (41%), moderate PA levels (42%), and high PA levels (42%), again considering the random effects of "Hypertension?" (red sections of Figure 4).

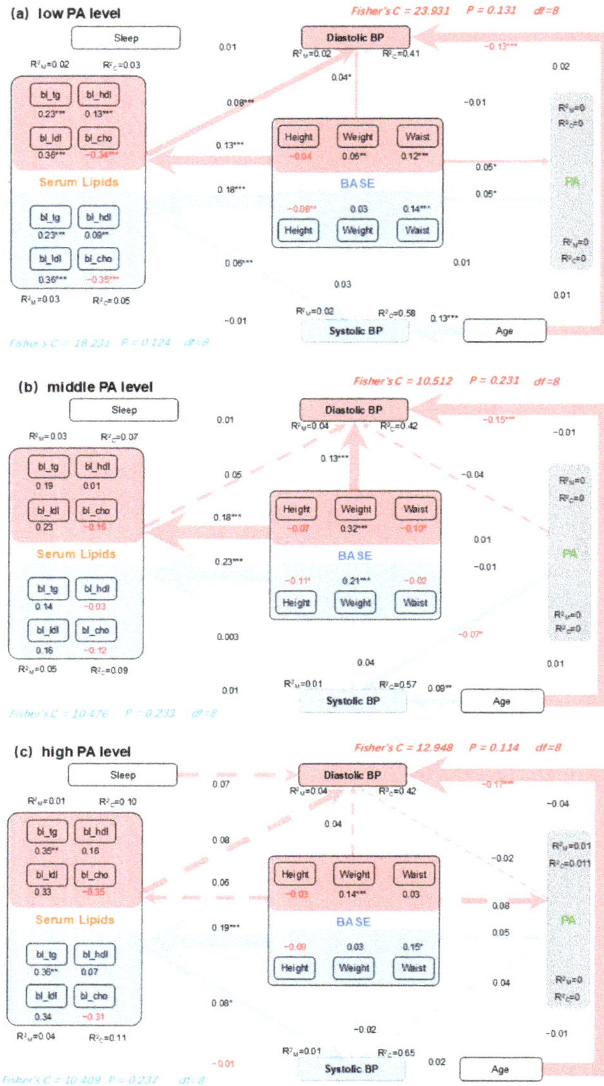

Figure 4. Piecewise SEM accounts for the direct and indirect effects of individual base, serum lipids, physical activity (PA), age, and sleep time on SBP (blue color) and DBP (red color) under low PA levels (**a**), moderate PA levels (**b**), and high PA levels (**c**), respectively. The individual base and serum lipids were combined into composite variables. Numbers adjacent to the measured variables represent their coefficients with the composite variables after the data were standardized. Numbers next to the arrows indicate path coefficients, which reflect the directly standardized effect size of the relationships. The thickness of the arrows represents the strength of these relationships. The total standardized effects of composite variables on blood pressure (BP) are shown in the marginal and conditional R^2, representing the proportion of variance explained by all predictors without (R^2m) and with (R^2c) accounting for random effects of "Hypertension?". Relationships between residual variables of measured predictors are not shown. Significance levels for each predictor are indicated as * $p < 0.05$, ** $p < 0.01$, and *** $p < 0.001$.

Aside from age, individual baseline variables—including height, weight, and waist circumference—consistently played a significant role in reflecting the regulations on both SBP and DBP, both directly and indirectly, by exerting influences on serum lipids and PA. Among these baseline variables, height and waist circumference were the most important factors affecting SBP (blue sections in Figure 4a,c) and DBP (red sections in Figure 4a,b) in response to changes in serum lipids. Weight significantly had impacts on serum lipids related to SBP, with HDL cholesterol (bl_hdl) being the most influential, particularly by reducing the effects at moderate PA levels (blue section in Figure 4b). Conversely, waist circumference had a strong impact on four serum lipids related to DBP by enhancing the effects at low PA levels (red section in Figure 4a). Furthermore, in both low and high PA level conditions, the mediating effects of serum lipids were particularly important for SBP (blue sections in Figure 4a,c), especially concerning triglycerides (bl_tg). Similarly, in low PA conditions, the mediating effects of serum lipids were also significant for DBP (red section in Figure 4a). Additionally, PA demonstrated only a weak mediating effect on DBP, particularly under low PA conditions (red section in Figure 4a). Moreover, we observed direct effects of serum lipids and PA on both SBP and DBP, which may also lead to indirect impacts from individual baseline variables across different PA levels (Figure 4).

4. Discussion

A lack of sufficient PA can negatively impact health in daily life, contributing to many chronic diseases, especially in the prevention of chronic diseases [17]. Our research revealed that PA levels could lead to significant differences in hypertension, based on ANOVA and logistic regression models. One previous meta-analysis found that individuals who engaged in moderate-intensity PA experienced an 18% reduction in the risk of developing hypertension [29]. A randomized clinical trial demonstrated that participants in the exercise group, who followed a 12-week aerobic program, experienced significant reductions in their 24 h ambulatory SBP and DBP [30]. These studies support our findings. There is already ample evidence confirming the effectiveness of sensible exercise in controlling hypertension. It is clear that appropriate forms of exercise and exercise intensity can contribute to the management and control of BP. Moreover, our findings suggest that achieving certain levels of PA intensity may be particularly important for hypertension management.

Serum lipids have been recognized as independent risk factors for atherosclerosis, coronary heart disease, and stroke. Our findings demonstrated that the four serum lipids are highly correlated with SBP and DBP, as well as with individual baseline variables. Among the continuous variables related to the risk of developing hypertension, we found that weight, age, and serum lipid-related variables were significantly associated with hypertension development ($p < 0.05$). Each unit increase in age and weight was associated with 1.43 times (OR = 1.43, 95% CI: 1.32–1.54, $p < 0.001$) and 1.84 times (OR = 1.84, 95% CI: 1.43–1.90, $p < 0.001$) increase in the risk of hypertension, respectively. These results are consistent with previous studies. The research by He et al. revealed that higher bl_tg and lower bl_hdl increased the risk of new-onset hypertension [31]. As age increases, blood vessels undergo degenerative changes, such as a decline in wall elasticity [32] and a deterioration in permeability [33], which prevents vessels from expanding and contracting efficiently, leading to increased BP. Previous research also indicated that several anthropometric indices show a significant correlation with hypertension. These indices can be included in an individual's medical history and used as tools for cardiovascular health screening, potentially yielding better results for public health [34]. These physiological processes related to aging are generally irreversible. Although age is an immutable risk factor for hypertension, adopting a healthy lifestyle may help minimize the risk of developing the condition. Our study also identified weight management as a key factor in preventing and controlling hypertension. A survey of Korean adolescents indicated that overweight and obese individuals had 1.52 and 1.89 times the risk of hypertension and exhibited a high clustering of cardiometabolic risks [35]. A study involving a Chinese population of middle-aged and elderly individuals revealed that all 14 indicators of obesity and adiposity were risk factors for hypertension [36]. Increased visceral lipids worsen lipid profiles and promote abdominal fat deposition, leading to inflammation and oxidative stress associated with obesity, atherosclerosis, and hypertension. These factors disrupt insulin metabolism and renal sodium retention, further affecting BP [36,37]. Our results revealed that all serum lipid variables were significantly associated with the incidence of hypertension. Specifically, triglycerides, high-density lipoprotein, and low-density lipoprotein were significant risk factors. In contrast, cholesterol was identified as a protective factor for hypertension. The mean cholesterol levels in both hypertensive and non-hypertensive elderly residents were within the recommended safe range (<200 mg/dL) in this study, which may explain the reason.

In addition, the impact of waist circumference on BP cannot be ignored. Evidence suggests that increased waist circumference elevates the risk of hypertension in older individuals [22]. Wang et al. analyzed data from a survey involving 5742 men and 5972 women, finding that waist circumference could potentially raise BP levels [23]. These studies highlight the significant influence of factors such as weight and waist circumference on BP. Our descriptive statistics demonstrate that hypertensive patients generally have higher values for all body morphology variables compared to non-hypertensive individuals, further underscoring the strong association between obesity and hypertension. The link between sleep and BP should also not be overlooked [27].

The underlying mechanisms of the relationship between obesity and hypertension are not fully understood. According to our SEM results, serum lipids exert the most significant mediating effect on BP, particularly under low PA conditions, while PA has a weaker mediating effect on BP, especially under high PA conditions. Additionally, height and waist circumference or weight are the most important variables influencing BP in response to increased serum lipids. Notably, high-density lipoprotein showed the most significant mediating effect in reducing the impact of weight on SBP under moderate PA conditions. It is well-known that high-density lipoprotein serves as the main apolipoprotein in the human body, promoting the synthesis of free cholesterol accumulated in peripheral tissues and lipoproteins in the bloodstream. It transports cholesterol to the liver for metabolism, maintaining lipid balance. As adipose tissue increases, HDL consumption rises, leading to the accumulation of unmetabolized neutral fats and lipid substances, which increases blood viscosity and contributes to dyslipidemia, thereby elevating blood pressure [13]. The relationship between high-density lipoprotein and BP is complex and varies under different conditions, indicating that more mechanistic studies at the cellular and molecular levels are needed to clarify this connection. Under both low and high PA conditions, triglycerides showed significant mediating effects on SBP. These findings suggest that serum lipids influence BP both directly and indirectly, while also highlighting the role of PA levels. Moreover, these proofs elucidate the associations among serum lipids, PA, and the primary CVD risk factors in elderly adults. They also focus on the internal relationships when PA and serum lipids do not show direct correlations and explore possible reasons through mediation effect analysis. Previous studies have indicated a positive association between weight and serum triglycerides [11], which is consistent with our findings, notably in relation to DBP under low PA conditions.

Our results suggest that increased lipid release from the body into the bloodstream under low PA conditions regulates metabolic pathways and influences BP. Although our findings did not show significant mediating effects of PA, some research indicates that serum lipids may act as moderators of PA [38]. The main reason for this could be the confounders adjusted in the model. We adjusted for all potential variables and random factors that may have contributed to the weak mediating effects of PA. Our results indicate that reducing adipose tissue is an important strategy for preventing and controlling hypertension at low PA levels. Additionally, increasing PA levels may affect blood lipid levels, and elevated fat could lead to hyperinsulinemia, disorders of the renin–angiotensin–aldosterone system, and ultimately vasoconstriction [39]. Previous studies have noted that dyslipidemia is closely related to hypertension [13]. This relationship is primarily reflected in excessive dietary intake of saturated fatty acids and a lack of PA, leading to increased blood viscosity, thickening of blood vessel walls, heightened mobility of vascular, and ultimately vascular lumen stenosis, reduced blood flow speed, and increased vascular pressure [23,39].

Based on mechanisms identified in previous studies, it is suggested that PA can influence BP. Furthermore, the impact of PA on the hypertension health of urban residents should not be ignored. Selecting appropriate intensity and types of PA can significantly affect BP levels. Although the results were verified through statistical analysis, this study has its limitations. The analysis of mediation variables may be insufficient. We adjusted for as many confounders as possible, and the estimated values were reasonable and strictly controlled. As more confounders are controlled, the stability of the relationships among independent variables, dependent variables, and mediating variables becomes contingent on sample size and data structure, necessitating the application of a multilevel mediation effect model. The anthropometric methods used to measure PA were not the most precise available, and a significant proportion of participants were excluded from our analysis. Therefore, generalizing our results to the entire survey population should be performed with caution. Furthermore, the current study has several limitations. The PA data came from self-reported questionnaires, and there may be some bias. The attrition of participants and the replenishment of queues are inevitable and may affect the results of the analysis. In future studies, we could consider additional variables and construct more reasonable frameworks using a multiple-level moderated mediation effect model to reveal more robust associations. Meanwhile, other microsystem indicators, such as metabolism-related indicators, hypertension-related genes or expressed proteins, and other molecular indicators, may effectively address the management and prevention of hypertension at various exercise intensities.

5. Conclusions

(1) Serum lipids have significant nonlinear relationships with BP, and PA levels have varying effects on BP. Meanwhile, the BP health of urban residents is influenced by factors such as age, weight, and waist measurements.

(2) There are significant mediating effects of serum lipids and weak mediating effects of PA on the relationship between individual body variables and SBP/DBP. These effects differ among PA levels, highlighting the importance of low PA levels in elderly hypertension management.

Author Contributions: F.W.: conceptualization, supervision, and funding acquisition. Y.Z. (Yang Zhao): writing—original draft and funding acquisition; Y.Z. (Yike Zhang): resources and writing—reviewing and editing. All authors have read and agreed to the published version of the manuscript.

Funding: This work was funded by financial support from Planning Subject for the 13th five-year plan of national education sciences (BLA200219), the Social Development Foundation of Shanxi, China (No. 201903D321069), the Applied Basic Research Project of Shanxi, China (No. 201801D121261), the Fund for Shanxi "Higher Education Teaching Reform and Innovation Project" (Nos. J20240124 and J20240195), Discipline Basic Capacity Building Project of Shanxi University (No. 2024-14), and "Research Project Supported by Shanxi Scholarship Council of China" (No. 2023-029).

Institutional Review Board Statement: The study was conducted in accordance with the Declaration of Helsinki, and approved by the Institutional Review Board of Peking University (protocol code IRB00001052-11015 and date of approval: 1 June 2008).

Informed Consent Statement: Informed consent was obtained from all subjects involved in the study.

Data Availability Statement: The data and R codes that support the findings of this study are available on request from the corresponding author.

Acknowledgments: This research uses data from the CHARLS. We thank all the participants in our study and the staff responsible for conducting CHARLS. We thank the reviewers' valuable comments.

Conflicts of Interest: The authors declare no conflicts of interest.

References

1. Boersma, P.; Black, L.I.; Ward, B.W. Prevalence of Multiple Chronic Conditions Among US Adults, 2018. *Prev. Chronic Dis.* **2020**, *17*, e106. [CrossRef]
2. Fuchs, F.D.; Whelton, P.K. High blood pressure and cardiovascular disease. *Hypertension* **2020**, *75*, 285–292. [CrossRef] [PubMed]
3. Wang, Z.; Chen, Z.; Zhang, L.; Wang, X.; Hao, G.; Zhang, Z.; Shao, L.; Tian, Y.; Dong, Y.; Zheng, C. Status of hypertension in China: Results from the China hypertension survey, 2012–2015. *Circulation* **2018**, *137*, 2344–2356. [CrossRef] [PubMed]
4. Wang, J.-G.; Zhang, W.; Li, Y.; Liu, L. Hypertension in China: Epidemiology and treatment initiatives. *Nat. Rev. Cardiol.* **2023**, *20*, 531–545. [CrossRef]
5. Xing, L.; Jing, L.; Tian, Y.; Lin, M.; Du, Z.; Yan, H.; Ren, G.; Dong, Y.; Sun, Q.; Liu, S. Urban–Rural disparities in status of hypertension in northeast China: A population-based study, 2017–2019. *Clin. Epidemiol.* **2019**, *11*, 801–820. [CrossRef] [PubMed]
6. NCD Risk Factor Collaboration (NCD-RisC). Worldwide trends in hypertension prevalence and progress in treatment and control from 1990 to 2019: A pooled analysis of 1201 population-representative studies with 104 million participants. *Lancet* **2021**, *398*, 957–980. [CrossRef] [PubMed]
7. Kearney, P.M.; Whelton, M.; Reynolds, K.; Muntner, P.; Whelton, P.K.; He, J. Global burden of hypertension: Analysis of worldwide data. *Lancet* **2005**, *365*, 217–223. [CrossRef] [PubMed]
8. Zhou, M.; Wang, H.; Zeng, X.; Yin, P.; Zhu, J.; Chen, W.; Li, X.; Wang, L.; Wang, L.; Liu, Y.; et al. Mortality, morbidity, and risk factors in China and its provinces, 1990–2017: A systematic analysis for the Global Burden of Disease Study 2017. *Lancet* **2019**, *394*, 1145–1158. [CrossRef] [PubMed]
9. GBD 2017 Risk Factor Collaborators. Global, regional, and national comparative risk assessment of 84 behavioural, environmental and occupational, and metabolic risks or clusters of risks for 195 coun-tries and territories, 1990–2017: A systematic analysis for the Global Burden of Disease Study 2017. *Lancet* **2018**, *392*, 1923–1994.
10. Khemka, S.; Reddy, A.; Garcia, R.I.; Jacobs, M.; Reddy, R.P.; Roghani, A.K.; Pattoor, V.; Basu, T.; Sehar, U.; Reddy, P.H. Role of diet and exercise in aging, Alzheimer's disease, and other chronic diseases. *Ageing Res. Rev.* **2023**, *91*, 102091. [CrossRef]
11. Mutalifu, M.; Zhao, Q.; Wang, Y.; Hamulati, X.; Wang, Y.-S.; Deng, L.; Adili, N.; Liu, F.; Yang, Y.-N.; Li, X.-M. Joint association of physical activity and diet quality with dyslipidemia: A cross-sectional study in Western China. *Lipids Health Dis.* **2024**, *23*, 46. [CrossRef]
12. Hedayatnia, M.; Asadi, Z.; Zare-Feyzabadi, R.; Yaghooti-Khorasani, M.; Ghazizadeh, H.; Ghaffarian-Zirak, R.; Nosrati-Tirkani, A.; Mohammadi-Bajgiran, M.; Rohban, M.; Sadabadi, F.; et al. Dyslipidemia and cardiovascular disease risk among the MASHAD study population. *Lipids Health Dis.* **2020**, *19*, 42. [CrossRef]
13. Dąbrowska, E.; Narkiewicz, K. Hypertension and dyslipidemia: The two partners in endothelium-related crime. *Curr. Atheroscler. Rep.* **2023**, *25*, 605–612. [CrossRef]
14. Booth, F.W.; Gordon, S.E.; Carlson, C.J.; Hamilton, M.T. Waging war on modern chronic diseases: Primary prevention through exercise biology. *J. Appl. Physiol.* **2000**, *88*, 774–787. [CrossRef] [PubMed]
15. Börjesson, M.; Onerup, A.; Lundqvist, S.; Dahlöf, B. Physical activity and exercise lower blood pressure in individuals with hypertension: Narrative review of 27 RCTs. *Br. J. Sports Med.* **2016**, *50*, 356–361. [CrossRef]
16. Leal, J.M.; Galliano, L.M.; Del Vecchio, F.B. Effectiveness of high-intensity interval training versus moderate-intensity continuous training in hypertensive patients: A systematic review and meta-analysis. *Curr. Hypertens. Rep.* **2020**, *22*, 26. [CrossRef]
17. Isath, A.; Koziol, K.J.; Martinez, M.W.; Garber, C.E.; Martinez, M.N.; Emery, M.S.; Baggish, A.L.; Naidu, S.S.; Lavie, C.J.; Arena, R. Exercise and cardiovascular health: A state-of-the-art review. *Prog. Cardiovasc. Dis.* **2023**, *79*, 44–52. [CrossRef] [PubMed]
18. Diaz, K.M.; Booth, J.N., III; Seals, S.R.; Abdalla, M.; Dubbert, P.M.; Sims, M.; Ladapo, J.A.; Redmond, N.; Muntner, P.; Shimbo, D. Physical activity and incident hypertension in African Americans: The Jackson Heart Study. *Hypertension* **2017**, *69*, 421–427. [CrossRef] [PubMed]
19. Byambasukh, O.; Snieder, H.; Corpeleijn, E. Relation between leisure time, commuting, and occupational physical activity with blood pressure in 125 402 adults: The lifelines cohort. *J. Am. Heart Assoc.* **2020**, *9*, e014313. [CrossRef] [PubMed]
20. Bull, F.C.; Al-Ansari, S.S.; Biddle, S.; Borodulin, K.; Buman, M.P.; Cardon, G.; Carty, C.; Chaput, J.-P.; Chastin, S.; Chou, R. World Health Organization 2020 guidelines on physical activity and sedentary behaviour. *Br. J. Sports Med.* **2020**, *54*, 1451–1462. [CrossRef]
21. Boakye, K.; Bovbjerg, M.; Schuna, J., Jr.; Branscum, A.; Varma, R.P.; Ismail, R.; Barbarash, O.; Dominguez, J.; Altuntas, Y.; Anjana, R.M. Urbanization and physical activity in the global Prospective Urban and Rural Epidemiology study. *Sci. Rep.* **2023**, *13*, 290. [CrossRef]
22. Sun, J.Y.; Ma, Y.X.; Liu, H.L.; Qu, Q.; Cheng, C.; Kong, X.Q.; Huang, W.J.; Sun, W. High waist circumference is a risk factor of new-onset hypertension: Evidence from the China Health and Retirement Longitudinal Study. *J. Clin. Hypertens.* **2022**, *24*, 320–328. [CrossRef]
23. Wang, F.; Zhang, X.; Wang, X.; Zhao, Y. Enhanced body shape change coupled with PA is the key to hypertension management for urban residents. *Public Health Nurs.* **2024**, *41*, 1016–1026. [CrossRef]
24. Zhao, Y.; Hu, Y.; Smith, J.P.; Strauss, J.; Yang, G. Cohort profile: The China health and retirement longitudinal study (CHARLS). *Int. J. Epidemiol.* **2014**, *43*, 61–68. [CrossRef]
25. Sember, V.; Meh, K.; Sorić, M.; Starc, G.; Rocha, P.; Jurak, G. Validity and reliability of international physical activity questionnaires for adults across EU countries: Systematic review and meta analysis. *Int. J. Environ. Res. Public Health* **2020**, *17*, 7161. [CrossRef]

26. Deng, Y.; Paul, D.R. The relationships between depressive symptoms, functional health status, physical activity, and the availability of recreational facilities: A rural-urban comparison in middle-aged and older Chinese adults. *Int. J Behav. Sci.* **2018**, *25*, 322–330. [CrossRef]
27. Zhang, J.; Zhang, F.; Xin, C.; Duan, Z.; Wei, J.; Zhang, X.; Han, S.; Niu, Z. Associations of long-term exposure to air pollution, physical activity with blood pressure and prevalence of hypertension: The China Health and Retirement Longitudinal Study. *Front. Public Health* **2023**, *11*, 1137118. [CrossRef]
28. Guthold, R.; Stevens, G.A.; Riley, L.M.; Bull, F.C. Worldwide trends in insufficient physical activity from 2001 to 2016: A pooled analysis of 358 population-based surveys with 1·9 million participants. *Lancet Glob. Health* **2018**, *6*, e1077–e1086. [CrossRef] [PubMed]
29. Xie, W.; Zhang, L.; Cheng, J.; Wang, Y.; Kang, H.; Gao, Y. Physical activity during pregnancy and the risk of gestational diabetes mellitus: A systematic review and dose–response meta-analysis. *BMC Public Health* **2024**, *24*, 594. [CrossRef] [PubMed]
30. Williamson, W.; Lewandowski, A.J.; Huckstep, O.J.; Lapidaire, W.; Ooms, A.; Tan, C.; Mohamed, A.; Alsharqi, M.; Bertagnolli, M.; Woodward, W. Effect of moderate to high intensity aerobic exercise on blood pressure in young adults: The TEPHRA open, two-arm, parallel superiority randomized clinical trial. *eClinicalMedicine* **2022**, *48*, 101445. [CrossRef]
31. He, D.; Fan, F.; Jia, J.; Jiang, Y.; Sun, P.; Wu, Z.; Li, J.; Huo, Y.; Zhang, Y. Lipid profiles and the risk of new-onset hypertension in a Chinese community-based cohort. *Nutr. Metab. Cardiovas. Dis.* **2021**, *31*, 911–920. [CrossRef] [PubMed]
32. Ungvari, Z.; Tarantini, S.; Kiss, T.; Wren, J.D.; Giles, C.B.; Griffin, C.T.; Murfee, W.L.; Pacher, P.; Csiszar, A. Endothelial dysfunction and angiogenesis impairment in the ageing vasculature. *Nat. Rev. Cardiol.* **2018**, *15*, 555–565. [CrossRef]
33. Sheikh, A.M.; Yano, S.; Tabassum, S.; Nagai, A. The Role of the Vascular System in Degenerative Diseases: Mechanisms and Implications. *Int. J. Mol. Sci.* **2024**, *25*, 2169. [CrossRef]
34. Feng, C.; Lu, C.; Chen, K.; Song, B.; Shan, Z.; Teng, W. Associations between various anthropometric indices and hypertension and hyperlipidaemia: A cross-sectional study in China. *BMC Public Health* **2024**, *24*, 3045. [CrossRef]
35. Han, Y.; Choi, Y.; Kim, Y.S. Association between physical activity levels and mortality across adiposity: A longitudinal study of age-specific Asian populations. *Geriatr. Gerontol. Int.* **2024**, *24*, 1156–1164. [CrossRef]
36. Wang, C.; Fu, W.; Cao, S.; Xu, H.; Tian, Q.; Gan, Y.; Guo, Y.; Yan, S.; Yan, F.; Yue, W. Association of adiposity indicators with hypertension among Chinese adults. *Nutr. Metab. Cardiovas.* **2021**, *31*, 1391–1400. [CrossRef]
37. Wen, X.; Zhang, B.; Wu, B.; Xiao, H.; Li, Z.; Li, R.; Xu, X.; Li, T. Signaling pathways in obesity: Mechanisms and therapeutic interventions. *Signal Transduct. Target. Ther.* **2022**, *7*, 298. [CrossRef] [PubMed]
38. Zou, Q.; Su, C.; Du, W.; Ouyang, Y.; Wang, H.; Zhang, B.; Luo, S.; Tan, T.; Chen, Y.; Zhong, X. The mediation and moderation effect association among physical activity, body-fat percentage, blood pressure, and serum lipids among Chinese adults: Findings from the China Health and Nutrition Surveys in 2015. *Nutrients* **2023**, *15*, 3113. [CrossRef]
39. Shams, E.; Kamalumpundi, V.; Peterson, J.; Gismondi, R.A.; Oigman, W.; de Gusmão Correia, M.L. Highlights of mechanisms and treatment of obesity-related hypertension. *J. Hum. Hypertens.* **2022**, *36*, 785–793. [CrossRef]

metabolites

MDPI

Article

Plasma Metabolomics Study on the Impact of Different CRF Levels on MetS Risk Factors

Xiaoxiao Fei, Qiqi Huang and Jiashi Lin *

College of physical Education, Jimei University, Xiamen 361021, China; fxiaoxiao1998@gmail.com (X.F.); qiqihuang072@gmail.com (Q.H.)
* Correspondence: linjiashi1980@gmail.com; Tel.: +86-134-5929-0239

Abstract: To investigate the metabolomic mechanisms by which changes in cardiorespiratory fitness (CRF) levels affect metabolic syndrome (MetS) risk factors and to provide a theoretical basis for the improvement of body metabolism via CRF in people with MetS risk factors, a comparative blood metabolomics study of individuals with varying levels of CRF and varying degrees of risk factors for MetS was conducted. Methods: Ninety subjects between the ages of 40 and 45 were enrolled, and they were categorized into low-MetS (LM $\leq$ two items) and high MetS (HM > three items) groups, as well as low- and high-CRF (LC, HC) and LCLM, LCLM, LCHM, and HCHM groups. Plasma was taken from the early morning abdominal venous blood. LC-MS was conducted using untargeted metabolomics technology, and the data were statistically and graphically evaluated using SPSS26.0 and R language. Results: (1) There were eight common differential metabolites in the HC vs. LC group as follows: methionine (↓), γ-aminobutyric acid (↑), 2-oxoglutatic acid (↑), arginine (↑), serine (↑), cis-aconitic acid (↑), glutamine (↓), and valine (↓); the HM vs. LM group are contrast. (2) In the HCHM vs. LCLM group, trends were observed in 2-oxoglutatic acid (↑), arginine (↑), serine (↑), cis-aconitic acid (↑), glutamine (↓), and valine (↓). (3) CRF and MetS risk factors jointly affect biological metabolic pathways such as arginine biosynthesis, TCA cycle, cysteine and methionine metabolism, glycine, serine, and threonine metabolism, arginine and proline metabolism, and alanine, aspartate, and glutamate metabolism. Conclusion: The eight common differential metabolites can serve as potential biomarkers for distinguishing individuals with different CRF levels and varying degrees of MetS risk factors. Increasing CRF levels may potentially mitigate MetS risk factors, as higher CRF levels are associated with reduced MetS risk.

Keywords: blood metabolomics; cardiorespiratory fitness; metabolic syndrome risk factors; metabolic pathways

1. Introduction

Metabolic syndrome (MetS) is characterized by a cluster of metabolic disorders, including hypertension, central obesity, dyslipidemia, and impaired glucose tolerance. Numerous studies have demonstrated that low cardiorespiratory fitness (CRF) is associated with an increased risk and mortality of MetS [1]. Both human and animal studies indicate that obesity, elevated fasting plasma glucose (FPG), and elevated insulin levels are common risk factors for MetS, particularly prevalent among individuals with low levels of CRF [2,3]. Further research is needed to elucidate the specific biochemical pathways influenced by high levels of CRF and their impact on MetS risk factors.

In recent years, there has been a growing emphasis among scholars on studying metabolic changes in the human body during exercise using metabolomics [4]. Methods such as nuclear magnetic resonance spectroscopy (NMR) and mass spectrometry (MS) enable the simultaneous measurement of a wide range of metabolites, revealing the biological remechanisms through which exercise improves health from a metabolic perspective, including its impact on MetS. Research over the past decade has primarily focused on comparing groups with significant disparities in CRF, such as those with the highest (>85%) and lowest (<15%) CRF levels or individuals with maximal oxygen consumption (VO_{2max}) ranging from 38 to 48 $mL^{-1}kg^{-1}$min and above 60 mL^{-1} kg^{-1}min [5,6]. This approach highlights that noticeable metabolic alterations are more evident when there are substantial differences in CRF levels. Current research has confirmed a link between CRF and several potential biomarkers related to MetS risk factors [7]. For instance, circulating levels of branched-chain amino acids (BCAAs) are significantly elevated in individuals affected by obesity, type 2 diabetes (T2D), and MetS compared to healthy controls [8]. For instance, there is evidence suggesting that circulating BCAA levels tend to decrease with increasing CRF levels or with improvements in health [9,10]. In a 16-week study, Sardeli et al. [11] combined aerobic and resistance exercise training for elderly women with MetS. Endurance training increased subjects' CRF levels by 131%, while resistance training enhanced the leg press resistance load from 48% to 160%. Notably, the reevaluation of MetS risk factors post-exercise showed no significant changes, while metabolomic analysis revealed a significant increase in substrates involved in the tricarboxylic acid (TCA) cycle, including 2-ketobutyric acid, ketone, and acetoacetate. This metabolic response may account for the improvement in fatty acid oxidation due to exercise. Exercise therapies have the potential to considerably improve CRF levels in MetS patients, which, in turn, can improve their metabolic status. Consequently, assessing a person's health only using their MetS risk factors may not be a reliable indicator of their general health.

This study employed High Performance Chemical Isotope Labeling (HP-CIL LC-MS) technology to conduct blood metabolomics analysis on populations with varying levels of CRF and MetS risk factors. The study aims to investigate the mechanisms through which CRF levels influence MetS risk factors by analyzing differences in blood metabolites among individuals with different CRF levels and degrees of MetS risk factors. Additionally, it seeks to provide insights into how enhancing CRF can potentially improve metabolic conditions in individuals with MetS risk factors, contributing to strategies for promoting health through exercise.

2. Materials and Methods

2.1. Study Subjects

A total of 100 participants aged 40–65 were recruited from Xiamen City Fujian Province, China. Inclusion in the study was based on meeting at least one of the following diagnostic criteria for MetS-related risk factors: (1)central obesity was defined as a waist circumference $\geq$ 90 cm for men and $\geq$85 cm for women; (2) fasting triglyceride (TG) level $\geq$ 1.7 mmol/L or under treatment; (3) high-density lipoprotein cholesterol (HDL-cholesterol) level $\leq$ 1.04 mmol/L or under treatment; (4) hypertension, defined as a systolic blood pressure (SBP) $\geq$ 130 mmHg or a diastolic blood pressure (DBP) $\leq$ 85 mmHg, or previously diagnosed and treated for hypertension; (5) hyperglycemia, defined as a fasting plasma glucose (FPG) level $\geq$ 6.1 mmol/L and/or 2-h postprandial glucose level $\leq$7.8 mmol/L, or previously diagnosed and treated for T2D. Subjects who did not complete the full trial or used vasoactive drugs (e.g., antihypertensive drugs, statins) or dietary supplements during the trial were excluded. The height, weight, waist circumference, and blood pressure of the subjects were measured (recorded to two decimal places). Fasting blood samples were collected after a 12-h fast for the measurement of FPG, TG, total cholesterol (TC), HDL-cholesterol, LDL-cholesterol, and homocysteine (Hcy). Measurements were performed using a Beckman Coulter automated biochemical analyzer (AU680, Beckman Coulter, Brea, CA, USA).

2.2. Cardiorespiratory Fitness Testing and Grouping

CRF was estimated using an indirect test to calculate the subjects VO_{2max}. The step test method involved a 3-min exercise duration, during which subjects stepped up and down to a metronome set at 120 beats/min. Each subject wore a heart rate monitor to continuously record their heart rate throughout the experiment. The formula to estimate

VO_{2max} is as follows: Step index = 100 × exercise time(s)/(2 × sum of heart rates in the 2nd, 3rd, and 4th 30-s intervals after stopping exercise).

For men: $VO_{2max}(mL\ kg^{-1}\ min^{-1}) = 0.2588 \times$ step index + 24.170.

For women: $VO_{2max}(mL\ kg^{-1}min^{-1}) = 0.1912 \times$ step index + 17.264.

MetS grouping: Subjects with ≤2 MetS risk factors were assigned to the low-risk group (LM) and those with ≥3 MetS risk factors were assigned to the high-risk group (HM), based on the severity of MetS risk factors. There were 41 subjects in the LM group and 49 in the HM group.

CRF grouping: Of the subjects who took the CRF test, 11 were unable to finish it or participate and 79 participants provided valid data for the collection. Using tertiles, CRF was split into three levels as follows: low (LC); medium (MC); and high (HC). The LC group consisted of 26 individuals; the MC group of 26; and the HC group of 27.

CRF + MetS grouping: Based on CRF levels and MetS risk factors, subjects were divided into four groups as follows: low CRF and low MetS risk (LCLM), low CRF and high MetS risk (LCHM), high CRF and low MetS risk (HCLM), and high CRF and high MetS risk (HCHM). There were 18 subjects in the LCLM group, 8 in the LCHM group, 12 in the HCLM group, and 15 in the HCHM group.(see Figure 1)

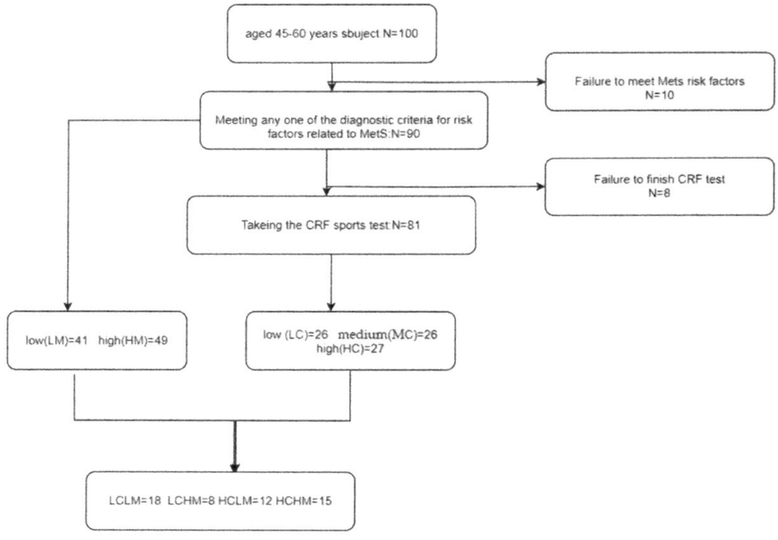

Figure 1. Flowchart of different metabolic syndrome MetS and cardiorespiratory fitness CRF groupings. Participants were categorized into the low-risk group and the high-risk group (HM) if they had ≥3 MetS-related risk factors; CRF was categorized into low CRF level (LC), medium CRF level (MC), and high CRF level (HC) using the 3-quartile method; participants were classified into four groups based on the CRF level and the number of MetS-related risk factors: low CRF and low MS risk factor group (LCLM), low CRF and high MS risk factor group (LCHM), high CRF and low MS risk factor group (HCLM), and high CRF and high MS risk factor group (HCHM).

2.3. Blood Sample Collection

Subjects fasted for 12 h and rested for 5 min to induce calmness. With their elbows elevated, 2 mL of venous blood were drawn into purple-top tubes containing anticoagulant. The tubes were gently inverted five to eight times, incubated at room temperature (20 to 25 °C) for 25–30 min, and then centrifuged at 3000 rpm for 10 min. The top layer of plasma was separated and stored in EP tubes pre-cooled to −80 °C for testing.

2.4. Metabolite Extraction

Plasma was divided into aliquots based on the requirements of the analysis. After thawing and vortexing, 30 μL of plasma from each sample was transferred into respective 1.5 mL centrifuge tubes. Each sample was divided into three parts as follows: one for single-channel analysis (30 μL per channel), one as a backup, and one for preparing mixed samples. For the mixed sample, 35 μL of plasma from each sample was pooled, carefully vortex-mixed to ensure uniformity, and labeled as a reference sample.

2.5. Protein Precipitation

To precipitate proteins, 90 μL of pre-cooled methanol (mass spectrometry-grade) was added to each centrifuge tube containing 30 μL of plasma. After thorough vortexing and centrifugation at low speed, the samples were placed in a −20 °C freezer for 1 h and subsequently centrifuged at high speed at 4 °C (12,000 rpm, 10 min). Ninety microliters of supernatant were transferred to new centrifuge tubes and dried under nitrogen.

2.6. Sample Labeling

For the secondary metabolomics analysis of amines and phenols, aliquoted samples were reconstituted by adding 25 μL of mass spectrometry-grade water. To each sample, 12.5 μL of buffer reagent A and 37.5 μL of ^{12}C labeling reagent B (for single and mixed sample labeling) or ^{13}C labeling reagent B (for mixed sample labeling only) were added. After vortex mixing, the samples were incubated at 40 °C for 45 min. To quench excess labeling reagent, 7.5 μL of reagent C was added post-incubation, followed by 30 μL of pH regulator D.

2.7. Sample Mixture

Following standard operating protocols, the labeled amine/phenolic secondary metabolome was analyzed using LC-UV. An aliquot of the ^{13}C-labeled mixed sample was combined with the ^{12}C-labeled individual sample for liquid-quality analysis, guided by the quantification results. The QC samples were prepared simultaneously with the liquid-quality analysis; specifically, equal volumes of mixed samples labeled with ^{13}C and ^{12}C were thoroughly mixed and used as QC samples. Each sample was prepared for analysis using liquid-quality analysis.

2.8. Analysis Condition and Data Quality Control and Metabolite Identification Results

LC-MS analysis was performed according to standardized operating procedures. Quality control and retention time calibration samples were analyzed every 15 injections to monitor instrument stability (see Table 1).

Table 1. UHPLC-Q-TOF/MS condition.

Name	Conditions of Association
chromatograph	Agilent 1290 Ultra High Performance Liquid Chromatography−6546 Quadrupole-Time of Flight Mass Spectrometer
column	Agilent eclipse plus reversed-phase C18 column (150 mm × 2.1 mm, 1.8 μm particle size)
mobile phase A	0.1% (*v*/*v*) Formic acid–water
mobile phase B	0.1% (*v*/*v*) Formic acid–acetonitrile
Gradient elution	t = 0 min, 25% MPB; t = 10 min, 99% MPB; t = 13 min,99% MPB t = 15 min, 99% MPB; t = 15.1 min, 25% MPB; t = 18 min, 25% MPB
Flow rate	400 μL/min
column temperature	40 °C
Scan range	*m*/*z* 220−1000 Da

Quality Accuracy Check: As depicted in Figure 2a,b, in the amine/phenol channel analysis, the chromatographic peak with a mass-to-charge ratio (*m*/*z*) of 251.0849 was chosen as the reference peak for assessing the quality accuracy of data from 97 samples.

Across all samples analyzed, the total signal intensity remained stable, and all scanned mass-to-charge ratios fell within the expected range, demonstrating the robust stability and accuracy of the collected data.

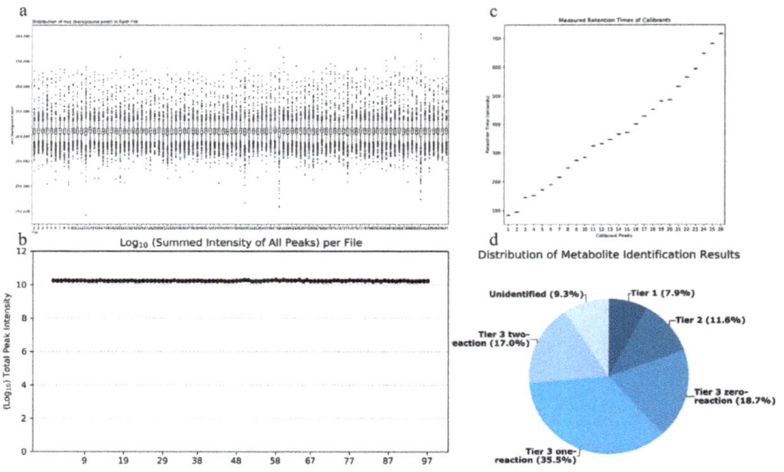

Figure 2. (**a**) Total signal intensity in the amine/phenol-based channel sub-sample analysis; (**b**) Background peak mass distribution in the amine/phenol-based channel sub-sample analysis; (**c**) Retention time assay results; (**d**) Distribution of identified metabolites in different tiers of the database.

Retention Time Check: As depicted in Figure 2c, data from 7 retention time calibration standards were utilized to verify the retention times. All standard peaks were well-resolved, and the retention times of the standards across all calibration data were consistent, indicating the robust stability of the collected data's retention times.

Metabolite Identification Results: Out of the detected metabolite peaks comprising 1851 features, 1679 peaks (90.7%) were accurately identified or inferred. As shown in Figure 2d, the three-tiered metabolite identification approach was applied as follows: Level 1, based on accurate molecular weight and retention time, identified 146 metabolites in the standard database (CIL Library); Level 2, utilizing the associated metabolite database (LI Library), identified an additional 214 metabolites; Level 3, where the remaining metabolites were matched in the MyCompoundID (MCID) database, resulting in matches for 347 metabolites in the 0-level reaction database, 657 metabolites in the 1-level reaction database, and 315 metabolites in the 2-level reaction database.

2.9. Statistical Analysis

The basic information of the subjects was statistically analyzed using SPSS 26.0 and GraphPad Prism 9.5 software. Results were expressed as mean ± SD, with statistically significant differences denoted by $p < 0.05$. The normality of the data was assessed using the Shapiro–Wilk test. The two independent samples t-test was applied when the data between two groups were normally distributed, and the non-parametric Mann–Whitney U-test was used for non-normal distribution. For comparisons involving three or more groups, one-way ANOVA was employed under the assumption of normal distribution and homogeneity of variances. Post hoc analysis was conducted using the LSD method for multiple comparisons, while the Kruskal–Wallis H-test was used for non-normally distributed data or when variances were not uniform.

Blood samples were analyzed using LC-MS metabolomics technology, and the data were processed and analyzed with Iso MS Pro 1.2.16 software. A three-tiered metabolite identification approach was used to structurally identify compounds and complete the initial metabolite data. The normalized data matrix was imported into SIMCA-p 14.1

software for multivariate data analysis, using partial least squares discriminant analysis (PLSA-DA) and orthogonal partial least squares-discrimination analysis (OPLS-DA) to compare differences in plasma metabolome. Regarding the variable importance projection (VIP) > 1.0 and nonparametric tests $p < 0.05$ as standard, choose differential metabolites. To prevent the overfitting of the OPLS-DA model, response ordering tests (200 random permutations) and cross-validation are used to evaluate the model's quality.

Following data processing, the fold change (FC) values for each group were determined using the sample mean ratio. (p-values from t-tests, q-values from FDR-adjusted p-values.) Differential metabolites between groups were selected based on FC values, p-values, and q-values obtained by volcano plot analysis, Venn diagram analysis, and VIP values. Metabolic pathway analysis was performed on the MetaboAnalyst 5.0 metabolomics analysis platform (https://www.metaboanalyst.ca/, accessed on 7 August 2022), and metabolic pathways were identified using the Kyoto Encyclopedia of Genes and Genomes (KEGG).

3. Results

3.1. Subjects

Subjects were classified by VO_{2max} into LC (28.90 ± 1.37, n = 26), MC (35.78 ± 3.41, n = 26), and HC (43.18 ± 1.46, n = 27) ($p < 0.05$). Compared to the LC group, the MC group had a significantly lower age ($p < 0.05$), while the HC group had significant differences in height, weight, WC, and HDL-cholesterol ($p < 0.05$). Compared to the MC group, HDL-cholesterol was significantly lower in the HC group ($p < 0.05$).

Subjects were divided into LM and HM groups based on the degree of MetS risk factors. Compared to the LM group, significant differences were observed in VO_{2max}, height, weight, BMI, WC, FPG, TG, TC, HDL-C, and Hcy in the HM group ($p < 0.05$) (see Table 2).

Table 2. Basic information of subjects in CRF and MetS groups.

Variables	CRF Group			MS Group	
	LC (n = 26)	MC (n = 26)	HC (n = 27)	LM (n = 49)	HM (n = 41)
Age (year)	54.19 ± 6.01	53.38 ± 6.33 *	52.85 ± 6.66	53.63 ± 5.98	52.17 ± 6.55
Height (cm)	160.94 ± 6.53	167 ± 8.09	168.63 ± 6.52 *	163.12 ± 6.71	168.63 ± 7.59 &
Weight (kg)	61.42 ± 9.01	67.91 ± 15.63	76.58 ± 24.03 *	61.26 ± 8.40	77.98 ± 20.88 &
BMI (kg/m^2)	23.63 ± 2.38	24.11 ± 3.89	26.85 ± 8.33	22.96 ± 2.15	27.30 ± 6.82 &
WC (cm)	80.30 ± 6.93	85.98 ± 13.05	88.56 ± 8.96 *	80.07 ± 7.00	90.08 ± 11.28 &
SBP (mmHg)	121.81 ± 17.90	126.69 ± 17.85	130.67 ± 17.70	122.45 ± 17.83	129.51 ± 15.87
DBP (mmHg)	79.42 ± 12.27	78.42 ± 12.81	82.85 ± 12.61	77.88 ± 12.75	82.34 ± 10.83
FPG (mmol/L)	5.39 ± 0.85	5.59 ± 1.43	5.26 ± 0.58	5.17 ± 0.73	5.67 ± 1.14 &
TG (mmol/L)	1.68 ± 1.07	1.65 ± 1.26	2.31 ± 2.00	1.30 ± 0.73	2.61 ± 1.78 &
TC (mmol/L)	5.33 ± 1.00	5.11 ± 1.49	5.01 ± 1.41	5.22 ± 1.25	4.99 ± 1.27 &
HDL-C (mmol/L)	1.21 ± 0.43	1.17 ± 0.33	0.92 ± 2.07 *#	1.27 ± 0.37	0.88 ± 0.13 &
LDL-C (mmol/L)	3.57 ± 0.85	3.33 ± 1.27	3.41 ± 1.28	3.50 ± 1.15	3.29 ± 1.04
Hcy (μmmol/L)	11.46 ± 4.35	13.41 ± 4.77	13.18 ± 2.33	11.90 ± 3.80	13.66 ± 4.10 &
VO_{2max}	28.90 ± 1.37	35.78 ± 3.41 *	43.18 ± 1.46 *#	34.54 ± 6.21	38.04 ± 5.92 &

In the CRF group, "*" for $p < 0.05$ compared with the LC group, and "#" for $p < 0.05$ compared with the MC group. In the MS group, "&" for $p < 0.05$ compared with the LM group.

A total of 53 subjects were included based on the CRF level and degree of MetS risk factors. Compared to the LCLM group, significant differences were observed in weight, BMI, WC, and HDL-cholesterol in the LCHM group ($p < 0.05$), in VO_{2max}, height, weight, and WC in the HCLM group ($p < 0.05$), and in VO_{2max}, height, weight, BMI, WC, SBP, TG, HDL-cholesterol, and Hcy in the HCHM group ($p < 0.05$). Compared to the LCHM group, significant differences were observed in VO_{2max} and TG in the HCLM group ($p < 0.05$) and in VO_{2max} and HDL-cholesterol in the HCHM group ($p < 0.05$). Compared to the HCLM group, significant differences were observed in WC, SBP, FPG, TC, and HDL-cholesterol in the HCHM group (see Table 3).

Table 3. Basic information of subjects in CRF + MetS groups.

Variables	CRF + MS Group			
	LCLM (n = 18)	LCHM (n = 8)	HCLM (n = 12)	HCHM (n = 15)
Age(year)	53.39 ± 6.01	56.00 ± 6.00	54.17 ± 6.71	51.80 ± 6.65
Height(cm)	158.78 ± 4.20	165.81 ± 8.38	168.26 ± 5.15 *	168.93 ± 7.61 *
Weight(kg)	57.98 ± 6.65	69.16 ± 9.16 *	67.72 ± 8.16 *	83.66 ± 29.98 *
BMI(kg/m²)	22.97 ± 2.07	25.13 ± 2.47 *	23.84 ± 1.82	29.25 ± 10.61 *
WC(cm)	78.54 ± 5.66	84.25 ± 8.24 *	84.68 ± 7.28 *	91.66 ± 9.18 *&
SBP(mmHg)	119.06 ± 18.59	128.00 ± 15.55	121.75 ± 17.92	137.80 ± 14.39 *&
DBP(mmHg)	79.17 ± 12.32	80.00 ± 13.00	77.67 ± 12.28	87.00 ± 11.63
FPG(mmol/L)	5.44 ± 0.99	5.28 ± 0.43	4.99 ± 0.30	5.47 ± 0.66 &
TG(mmol/L)	1.44 ± 1.03	2.21 ± 1.02	1.25 ± 0.36 #	3.15 ± 2.37 *&
TC(mmol/L)	5.56 ± 0.91	4.81 ± 1.06	4.95 ± 1.73	5.05 ± 1.16
HDL-C(mmol/L)	1.33 ± 0.46	0.93 ± 0.08 *	1.07 ± 0.23	0.82 ± 0.94 *#&
LDL-C(mmol/L)	3.74 ± 0.74	3.20 ± 1.01	3.50 ± 1.77	3.34 ± 0.77
Hcy(μmmol/L)	10.74 ± 3.59	13.09 ± 5.67	13.03 ± 3.26	13.30 ± 1.31 *
VO_{2max}	28.90 ± 1.38	28.90 ± 1.24	43.30 ± 1.71 *#	43.08 ± 1.29 *#

In the CRF + MS group, $p < 0.05$ compared with the LCLM group is "*"; $p < 0.05$ compared with the LCHM group is "#"; and $p < 0.05$ compared with the HCLM group is "&".

3.2. PLS-DA

Figure 3a shows the PLS-DA score plot of plasma metabolites for subjects with different CRF levels. Studies show that there is overlap between the plasma samples from the MC and HC groups, as well as between the LC and MC groups, making it impossible to identify between the three subject groups' plasma samples.

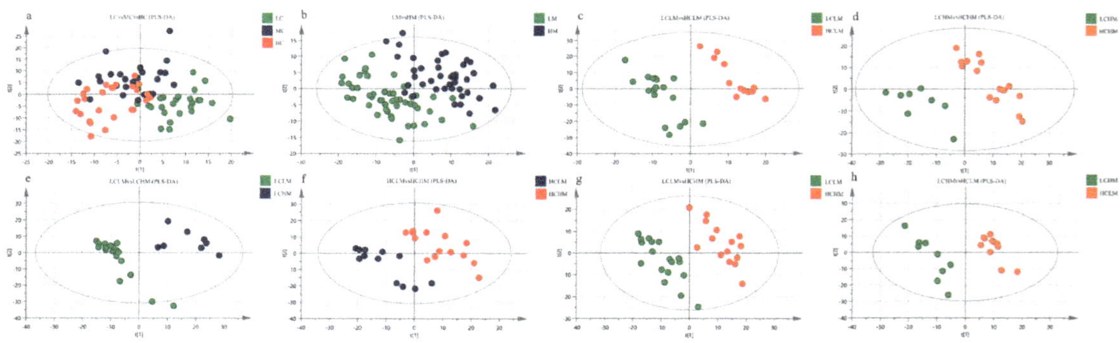

Figure 3. PLS-DA Score Plot: (**a**) LC vs. MC vs. HC group; (**b**) LM vs. HM group (**c**) LCLM vs. HCLM; (**d**) LCHM vs. HCHM; (**e**) LCLM vs. LCHM; (**f**) HCLM vs. HCHM; (**g**) LCLM vs. HCHM; (**h**) LCHM vs. HCLM.

The PLS-DA score plot of plasma metabolites for subjects with varying degrees of MetS risk factors is displayed in Figure 3b. The findings suggest that although there is a trend of separation, the plasma samples from the LM and HM groups cannot be fully differentiated from one another.

The PLS-DA score plots for the comparisons of the LCLM vs. HCLM, LCHM vs. HCHM, LCLM vs. HCHM, and LCHM vs. HCLM groups are shown in Figure 3c–e and Figure 3g,h, respectively. The findings demonstrate the total plasma sample separation, with no overlap, among these five comparison groups. The comparison between the HCLM and HCHM groups is depicted in Figure 3f, where the PLS-DA score plot indicates an imperfect separation of plasma samples between these groups with overlap.

The cross-validation results for each group are as follows: as shown in Table 4, the models of LC vs.MC vs. HC groups have relatively small values of R^2X, R^2Y, and Q^2, indicating that this model lacks reliability and predictability. $Q^2 < 0.5$ (HCLM vs. HCHM and LCHM vs. HCLM), indicating that the model is valid but lacks predictability. Models of LM vs. HM, LCLM vs. HCLM, LCHM vs. HCHM, LCLM vs. LCHM, and LCLM vs. HCHM groups demonstrate reliability and predictability.

Table 4. PLS-DA cross-validation parameters.

Group	R^2X	R^2Y	Q^2
LC vs. MC vs. HC	0.144	0.481	−0.063
LM vs.HM	0.322	0.967	0.520
LCLM vs. HCLM	0.249	0.981	0.557
LCHM vs. HCHM	0.377	0.992	0.625
LCLM vs. LCHM	0.365	0.992	0.715
HCLM vs. HCHM	0.290	0.946	0.195
LCLM vs. HCHM	0.368	0.997	0.813
LCHM vs. HCLM	0.333	0.998	0.296

R^2X denotes the explanatory rate of the model to the X matrix; R^2Y denotes the explanatory rate of the model to the Y matrix; and Q^2 denotes the predictive power of the model. The closer the three models are to 1, the better, and $Q^2 > 0.5$ is accepted.

3.3. OPLS-DA

Filtering out the irrelevant orthogonal signals by OPLS-DA analysis made the differences between groups more obvious and the differential metabolites obtained more reliable. OPLS-DA was used to analyze each differential comparison group of the CRF, MS, and CRF + MS groups and to verify whether the model was overfitted with cross-validation and response ranking

The metabolic status of the subjects in both groups was determined by the complete separation of sample points in the LC vs. HC group (see Figure 4A,a); the results show that the model is valid and there is no overfitting ($R^2X = 0.259$, $R^2Y = 0.974$, $Q^2 = 0.607$). The separation trend in plasma sample points was less distinct in both the MC vs. HC and LC vs. MC groups when compared to the LC vs. HC group(see Figure 4B,C,b,c). This suggests a less pronounced metabolic status in these groups ($R^2X = 0.162$, $R^2Y = 0.930$, $Q^2 = 0.310$; $R^2X = 0.139$, $R^2Y = 0.810$, $Q^2 = -0.152$); the models were not overfitted, as demonstrated by the response order tests, but the cross-validation results revealed that $Q^2 < 0.5$ for the two comparison groups, indicating that the models' predictability was lacking. The MS group's plasma isolation sample points in the LM vs. HM group were completely separated, and the model ($R^2X = 0.415$, $R^2Y = 0.998$, $Q^2 = 0.520$) was observable (see Figure 4D,d). The model's validity and predictability are all demonstrated by the cross-validation results, and these qualities are further demonstrated by the outcomes of the matching ranking test.

The OPLS-DA scores for each group in the CRF + MS group were as follows(see Table 5): (1) The sample points of the four groups—LCLM vs. HCLM, LCHM vs. HCHM, LCLM vs. HCHM, and LCLM vs. HCHM—were completely segregated, indicating significant differences in the participants' metabolic states (see Figure 4E–G,I,e–g,i). The results of the 200-response sequencing test confirmed the validity and predictability of the models in all four groups, and the cross-validation results showed their reliability ($R^2X = 0.296$, $R^2Y = 0.994$, $Q^2 = 0.548$; $R^2X = 0.262$, $R^2Y = 0.979$, $Q^2 = 0.520$; $R^2X = 0.282$, $R^2Y = 0.966$, $Q^2 = 0.613$; $R^2X = 0.342$, $R^2Y = 0.997$, $Q^2 = 0.651$). (2) The HCLM vs. HCHM group and the LCHM vs. LCHM group have completely separated sample points, and the metabolic differences between the subjects are obvious(see Figure 4H,J,h,j), which makes the model valid and reliable, but less predictive ($R^2X = 0.321$, $R^2Y = 0.989$, $Q^2 = 0.028$; $R^2X = 0.192$, $R^2Y = 0.958$, $Q^2 = 0.459$).

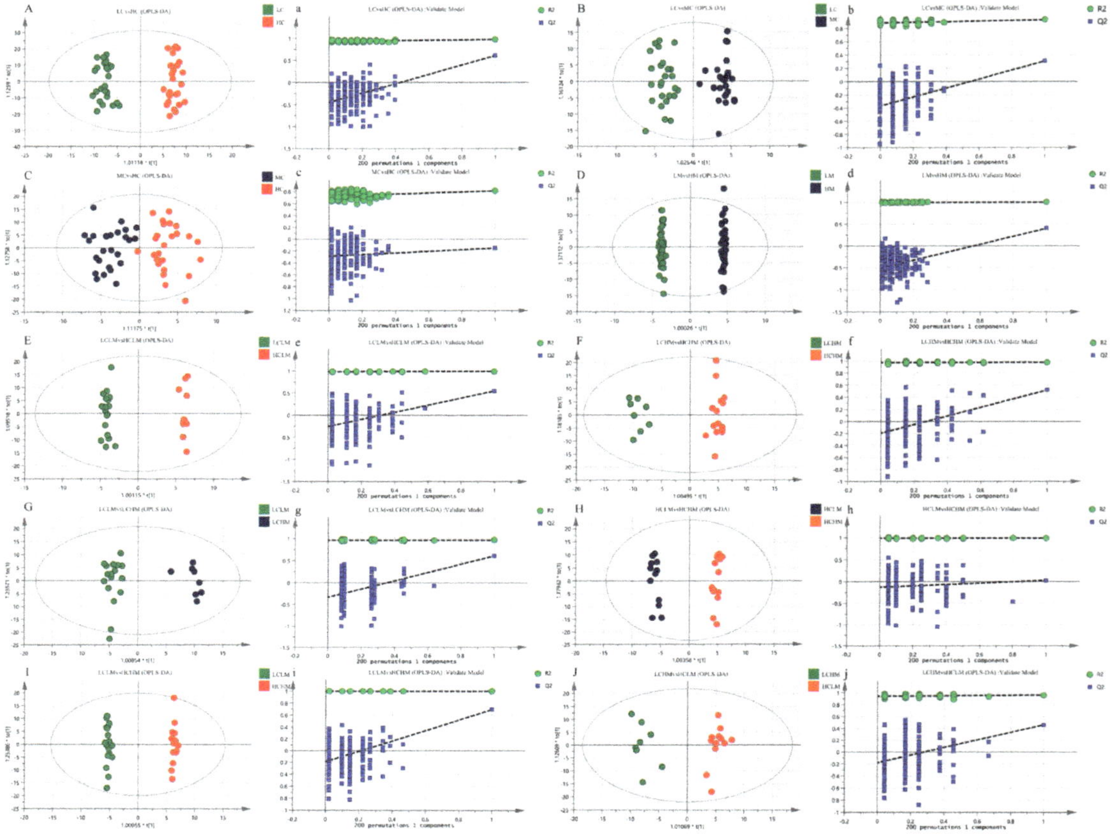

Figure 4. The OPLS-DA score plots (**A–J**) depict the scores for each group, while the response ranking test plots (**a–j**) display the results of the response ranking test. In the ranking test, the horizontal axis represents the correlation between the *Y* values of the random grouping and the *Y* values of the original grouping, while the vertical axis represents the R^2 and Q^2 scores.

Table 5. OPLS-DA cross-validation parameters.

Group	R^2X	R^2Y	Q^2
LC vs. HC	0.259	0.974	0.607
LC vs. MC	0.162	0.930	0.310
MC vs. HC	0.139	0.810	−0.152
LM vs. HM	0.415	0.998	0.520
LCLM vs. HCLM	0.296	0.994	0.548
LCHM vs. HCHM	0.262	0.979	0.520
LCLM vs. LCHM	0.287	0.966	0.613
HCLM vs. HCHM	0.321	0.989	0.028
LCLM vs. HCHM	0.342	0.997	0.651
LCHM vs. HCLM	0.192	0.958	0.459

The metabolism analysis demonstrated that each group's samples were well-clustered, while there was a significant separation between the groups. R^2Y and Q^2 values were computed in the OPLS-DA modeling study to demonstrate the model's validity. Q^2 values were also used to demonstrate the model's validity. A 200 random permutation test was conducted to demonstrate that the model did not overfit ($Q^2 < 0$) and to validate its reliability.

3.4. Volcano Plot Analysis

Given the results of the OPLS-DA analysis, volcano plot analyses were performed for the LC group vs. the HC group, the LM group vs. the HM group, the LCLM group vs. the HCLM group, the LCHM group vs. the HCHM group, the LCLM group vs. the LCHM group, and the LCLM group vs. the HCHM group. As can be seen in Figure 5a, 83 metabolites were upregulated and 16 metabolites were downregulated in the HC vs. LC group comparison; in Figure 5b, 84 metabolites were upregulated and 78 metabolites were downregulated in the HM vs. LM group comparison; in Figure 5c, 40 metabolites were upregulated and 11 metabolites were downregulated in the HCLM vs. LCLM group comparison; in Figure 5d, 121 metabolites were upregulated and 106 metabolites were downregulated in the HCHM vs. LCHM group comparison; in Figure 5e, 112 metabolites were upregulated and 99 metabolites were downregulated in the LCHM vs. LCLM group comparison; and in Figure 5f, 140 metabolites were upregulated and 61 metabolites were downregulated in the HCHM vs. LCLM group comparison.

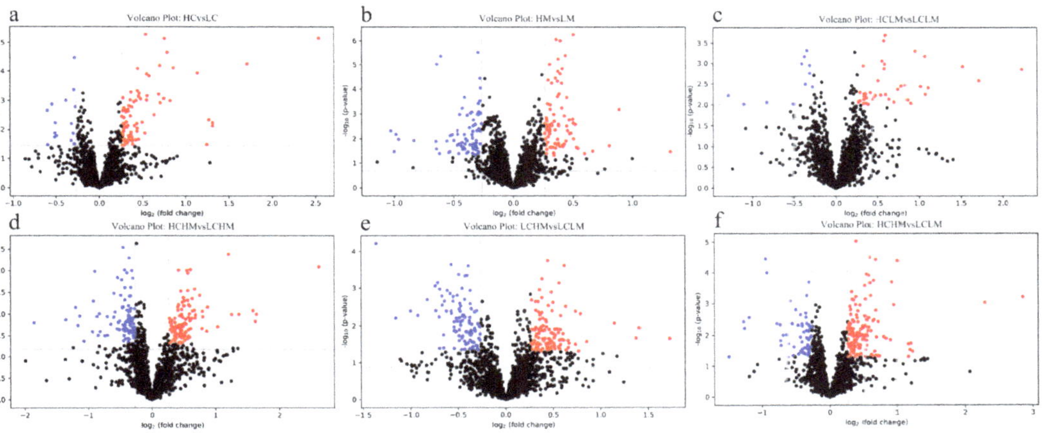

Figure 5. Differential metabolite volcano plot (**a**) HC vs. LC, (**b**) HM vs. LM, (**c**) HCLM vs. LCLM, (**d**) HCHM vs. LCHM, (**e**) LCHM vs. LCLM, (**f**) HCHM vs. LCLM. The horizontal axis represents the fold change in metabolite expression across different subgroups [log2(FoldChange)], while the vertical axis indicates the significance level of differences [−log10 (p-value)]. Each point on the plot represents a metabolite, with red indicating a significant increase, blue indicating a significant decrease, and black indicating no significant difference. Plasma metabolites were visualized on a volcano plot based on their fold change (FC) values, p-values, and q-values. VIP value indicates the contribution of each variable to the PLS-DA model. Differential metabolites were identified based on VIP values > 1, fold change (FC) > 1.2 or <0.83, $p < 0.05$, and $q < 0.25$.

3.5. Venn Diagram Analysis

Eight common differentially expressed metabolites were identified after excluding metabolites that could not be matched or identified. These eight frequently occurring differentially expressed metabolites may serve as biomarkers for distinguishing different levels of MetS risk factors and CRF levels. Methionine, γ-aminobutyric acid, 2-oxoglutaric acid, arginine, serine, cis-aconitic acid, glutamine, and valine showed significant differences in both comparison groups.

In the comparisons, 2-oxoglutaric acid and cis-aconitic acid were screened in the HCLM vs. LCLM group; methionine, serine, 2-oxoglutaric acid, arginine, and glutamine in the HCHM vs. LCHM group; and glutamine, methionine, and arginine in the LCHM vs.

LCLM group. Additionally, 2-oxoglutaric acid, arginine, serine, cis-aconitic acid, glutamine, and valine were screened in the HCHM vs. LCLM group (see Figure 6).

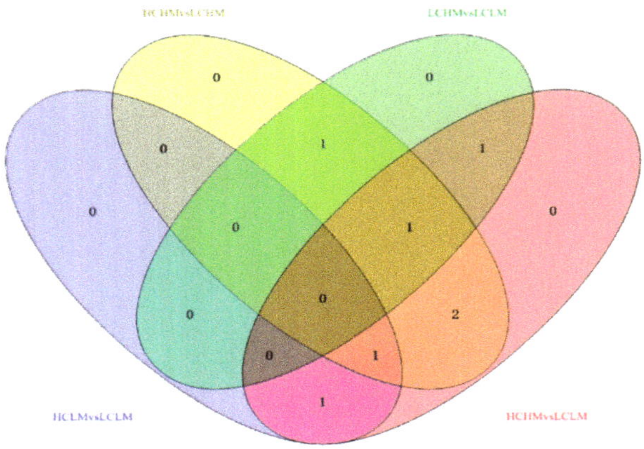

Figure 6. Venn diagram analysis of potential biomarkers in different groups.

The trends in the common differential metabolites in the comparison groups were different, as shown in Figure 6 and Table 6: (1) methionine: HC vs. LC and HCHM vs. LCHM (↓), HM vs. LM and LCHM vs. LCHM (↑); (2) γ-aminobutyric acid: HC vs. LC (↑), HM vs. LM (↓); (3) 2-oxoglutaric acid: HC vs. LC (↑), HM vs. LM (↓); HCLM vs. LCLM, HCHM vs. LCHM, and HCHM vs. LCLM (↑) (4) arginine: HC vs. LC (↑), HM vs. LM (↓), HCHM vs. LCHM, HCHM vs. LCHM (↑), LCHM vs. LCLM (↓) (5) serine: HC vs. LC (↑), HM vs. LM (↓), HCHM vs. LCLM (↑), LCHM vs. LCHM (↓); (6) cis-aconitic acid: HC vs. LC (↑), HM vs. LM (↓), HCLM vs. LCLM and HCHM vs. LCLM (↑); (7) glutamine: HC vs. LC (↓), HM vs. LM (↑), HCHM VS.LCHM and HCHM vs. LCLM (↓); (8) Valine: HC vs. LC (↓), HM vs. LM (↑); HCHM vs. LCHM and HCHM vs. LCLM (↓).

Table 6. Common differential metabolite FC values for each differential comparison group.

Compounds	CRF Group	MS Group	CRF + MS Group			
	HC vs. LC	HM vs. LM	HCLM vs. LCLM	HCHM vs. LCHM	LCHM vs. LCLM	HCHM vs. LCLM
L-Methionine	0.772 *	1.446 **	-	0.457 *	2.214 *	-
γ-Aminobutyric acid	1.284 *	0.701 *	-	-	-	-
2-Oxoglutaric acid	2.204 **	0.824 *	2.872 **	1.481 *	-	1.784 *
L-Arginine	1.343 *	0.787 *	-	1.251	0.746 **	1.429 **
L-Serine	1.268 *	0.725 **	-	-	0.703 **	1.404 **
cis-Aconitic acid	1.681 **	0.809 **	2.026 **	-	-	1.289 *
L-Glutamine	0.581 *	1.227 *	-	0.783 *	-	0.824 *
L-Valine	0.804 *	1.419 **	-	0.787 *	-	0.759 *

FC values are the multiplicity of differences, "*" indicating differential metabolite FC > 1.2 or <0.83, $p < 0.05$, and $q < 0.25$; "**" indicating differential metabolite FC > 1.2 or <0.83, $p < 0.01$, and $q < 0.10$.

Six common differentially expressed metabolites were identified across three comparison groups as follows: HM vs. LM, HC vs. LC, and HCHM vs. LCLM. As illustrated in Figure 7a–d, the following were found to be upregulated in the HC vs. LC and HCHM vs. LCLM group: 2-oxoglutaric acid, arginine, serine, and cis-aconitic acid; however, these same variables were found to be downregulated in the HM vs. LM group; similarly, Figure 7e,f revealed that valine and glutamine were both downregulated in the HCHM vs. the LCLM group and the HC vs. the LC group, but upregulated in the HM vs. the LCLM group.

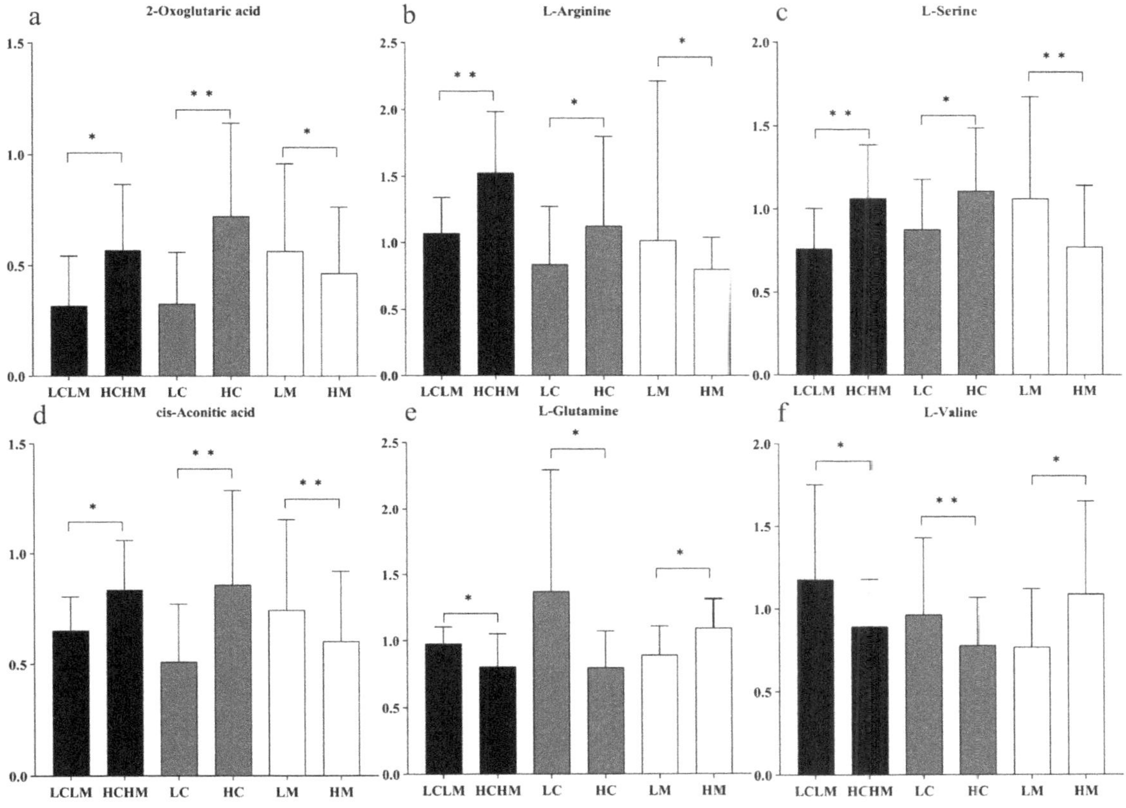

Figure 7. Histogram of trends in common differential metabolites by group. "*" indicating differential metabolite FC > 1.2 or <0.83, $p < 0.05$, and $q < 0.25$; "**" indicating differential metabolite FC > 1.2 or <0.83, $p < 0.01$, and $q < 0.10$; black column: HCHM group vs. LCLM group; grey column: HC group vs. LC group; white column: HM group vs. LM group. (**a**) 2-oxoglutaric acid; (**b**) L-arginine; (**c**) L-serine; (**d**) cis-aconitic acid; (**e**) L-glutamine; (**f**) L-valine.

3.6. Metabolite Pathway Analysis

For the purpose of determining the primary metabolic pathways impacting each group's metabolic status, Metaboanalyst 5.0 was also used for the topological analysis and enrichment of metabolic pathways of differential metabolites in each comparison group.

As illustrated in Figure 8a,b, the metabolic pathways of arginine biosynthesis, the TCA cycle, cysteine and methionine metabolism, glycine, serine, and threonine metabolism, arginine and proline metabolism, and alanine, aspartate, and glutamate metabolism were the metabolic pathways that were jointly affected by the HC vs. LC group and the HM vs. LM group.

The metabolic pathways of arginine biosynthesis, arginine and proline metabolism, as well as alanine, aspartate, and glutamate metabolism were jointly affected by the HC vs. LCLM group, the HCHM vs. LCHM group, the LCHM vs. LCLM group, and the HCHM vs. LCLM group, as shown in Figure 8c–f.

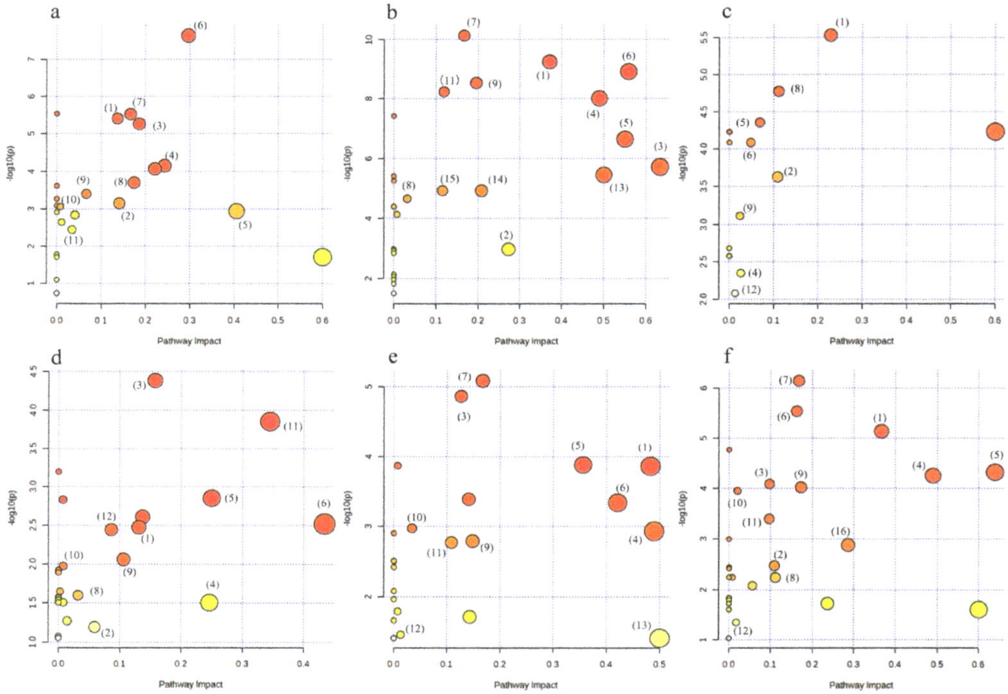

Figure 8. Bubble diagram of metabolic pathways of differential metabolites in each group. (**a–f**) represent the metabolic pathway bubble diagrams for each group, (**a**) HC vs. LC group; (**b**) HM vs. LM group; (**c**) HCLM vs. LCLM group; (**d**) HCHM vs. LCHM group; (**e**) LCHM group vs. LCLM group (**f**) HCHM vs. LCLM group (1): arginine biosynthesis; (2): tricarboxylic acid cycle(TCA); (3): cysteine and methionine metabolism; (4): glycine, serine, and threonine metabolism; (5): arginine and proline metabolism; (6): alanine, aspartic acid, and glutamate metabolism; (7): aminotransferase-tRNA biosynthesis; (8): Butyric acid metabolism; (9): glyoxylate and dicarboxylic acid metabolism; (10): purine metabolism; (11): glutathione metabolism; (12): glycerophospholipid metabolism; (13): D-glutamine and D-glutamate metabolism; (14): pyruvate metabolism; (15): glycolysis/glycohydrogenation; (16): taurine and hypotaurine metabolism. The graph's *X*-axis represents the pathway influence factor, and the *Y*-axis shows the enrichment analysis's *p*-value. The bigger the circle, the more influential it is; the darker the circle's color, the lower the *p*-value and the more significant the enrichment.

4. Discussion

This study aimed to investigate how variations in CRF levels impact MetS risk factors by conducting a metabolomics analysis on individuals with varying degrees of MetS. The study's primary conclusions were as follows: (1) By comparing intergroup differences between the HM vs. LM and the LC vs. HC, we identified eight common differential metabolites that serve as potential biomarkers for distinguishing between different CRF levels and degrees of MetS risk factors as follows: methionine, γ-aminobutyric acid, 2-oxoglutaric acid, arginine, serine, cis-aconitic acid, glutamine, and valine. (2) Six common differential metabolites, (methionine, γ-aminobutyric acid, 2-oxoglutaric acid, arginine, serine, cis-aconitic acid, glutamine, and valine) were found after the additional comparative analyses of the results of differential metabolites screened in the HCHM vs. LCLM, HC vs. LC, and HM vs. LM. These metabolites primarily affect arginine biosynthesis, the tricarboxylic acid (TCA) cycle, cysteine and methionine metabolism, glycine, serine and threonine metabolism, arginine and proline metabolism, and alanine, aspartate, and glutamate metabolism, among other biometabolic pathways.

The study identified how CRF levels influence metabolic status across varying degrees of MetS through metabolomic analysis, shedding light on CRF's impact on MetS risk via specific metabolic pathways. Changes in these metabolites may correlate with diverse CRF levels and MetS risk, offering novel perspectives for clinical diagnosis and intervention. The findings indicate that higher CRF levels can alleviate the detrimental effects of MetS risk factors, suggesting that enhancing fitness could effectively prevent or treat MetS. Moreover, the identification of six common differential metabolites impacting multiple vital biometabolic pathways enhances our comprehension of metabolic network complexity and proposes potential targets for intervention.

4.1. Differences in Plasma Metabolites between CRF Levels and Different Degrees of Mets Risk Factors

There are eight common differentially expressed metabolites in the HC vs. LC and HM vs. LM groups, namely methionine, γ-aminobutyric acid, 2-oxoglutaric acid, arginine, serine, aconitic acid, glutamine, and valine. They are potential biomarkers for distinguishing different CRF levels and degrees of MetS risk factors. Methionine is one of the essential amino acids in the human body and serves as an important cellular antioxidant [12]. It also participates in the regulation of insulin sensitivity. The present study observed a tendency for methionine downregulation in individuals with high CRF levels when compared to those with low CRF levels. Conversely, there was a tendency for methionine upregulation in subjects with high MetS risk factors compared to those with low MetS risk factors. Because exercise-induced oxidative stress inhibits methionine, both long-term endurance exercise can result in a substantial drop in plasma methionine levels [13]. Oxidative stress (OS) results from an imbalance between reactive oxygen species (ROS) production and antioxidant mechanisms. Regular moderate-intensity exercise has been demonstrated to enhance antioxidant defense mechanisms, thereby reducing mitochondrial oxidative damage [14,15]. However, the specific mechanism underlying the relationship between methionine depletion and its impact on oxidative stress induced by endurance exercise remains unclear. Past studies have demonstrated a positive association between blood levels of branched-chain and aromatic amino acids and MetS risk. However, the association of other essential amino acids such as methionine, threonine, and lysine with MetS remains poorly understood [16]. Azab et al. [17] discovered a favorable correlation between the prevalence of MetS and plasma levels of Met, Threonine, and Lysine. γ-aminobutyric acid (GABA) is generated from glutamate through the catalysis of glutamate decarboxylase. In this study, we observed significantly higher plasma levels of GABA in subjects with high CRF levels compared to those with low CRF levels. Conversely, plasma levels of GABA were significantly lower in subjects with high MetS risk factors compared to those with low MS risk factors. Yan et al. [18] found that, among male college students aged 18 to 29, serum GABA levels were significantly higher in groups with regular endurance exercise and strength exercise habits compared to those without such habits. Additionally, the GABA content of the endurance exercise habit group was significantly higher than that of the strength exercise habit group. These findings imply that prolonged endurance training will raise the level of GABA in the blood. Moreover, GABA modulated high-fat-induced obesity, glucose intolerance, and insulin resistance (IR). Oral GABA administration was shown to prevent weight gain, significantly lower FPG levels, improve glucose tolerance, and subsequently increase insulin sensitivity in high-fat-induced mice. Additionally, GABA significantly increased the number of regulatory T cells, which can control the development of obesity-induced IR and T2D [19]. Therefore, increasing CRF levels bringing about the upregulation of GABA content may be very favorable for intervening obesity and T2D occurrence.

Both Met and GABA can serve as potential biomarkers to distinguish CRF levels and the degree of MetS risk factors. Whether they can explain the mechanism by which high CRF levels improve the metabolism of individuals with MetS risk factors still needs further investigation.

4.2. Higher CRF Levels Reduce the Risk of MetS Risk Factors

In order to investigate the metabolomics mechanisms by which high levels of CRF affect high MetS risk factors, the interaction between CRF and MS revealed that high CRF levels, acting as a protective factor against high MetS risk, correlated with six metabolites. Four metabolites exhibited a downregulation trend in HM and an upregulation trend in HCHM. Conversely, two metabolites showed an upregulation in the HM and a downregulation in HCHM. The above metabolite trends indicate that the metabolite levels in the high MetS risk factor group were reversed due to the effect of high levels of CRF, which suggests that high levels of CRF can mitigate the detrimental effects of high MetS risk factors and contribute to improved metabolic status compared to individuals with low MetS risk factors.

α-Ketoglutaric acid is another name for ketoglutaric acid (Akg), an intermediary of the TCA cycle that is essential to several metabolic activities. We found a trend towards the upregulation of 2-oxoglutarate in the plasma of subjects with high levels of CRF compared to subjects with low levels of CRF, and a trend towards the downregulation of 2-oxoglutarate in subjects with high MS risk factors compared to subjects with low MS risk factors. Yuan et al. [20] examined peripheral blood samples from mice that were either inactive control mice or mice that had undergone acute resistance exercise (40 min of ladder climbing with a 10% body weight load). It was discovered that Akg concentrations were negatively correlated with HbA1c and several metabolic risk factors, including BMI, WC, adiposity, and body weight [21,22]. This helps to explain why patients with diabetes mellitus (DM) had significantly lower levels of Akg than healthy controls. But, when the HCHM and LCLM groups were compared, the trend in Akg increased rather than decreased, even though the HCHM group had a much higher weight, BMI, and WC than the LCLM group. The reason for this phenomenon could be that those in the HCHM group who have high CRF levels have already somewhat changed the metabolic status of people who have high MetS risk factors. Leibowitz et al. [23] found that exercise caused a significant increase in the amount of Akg in the blood, and that 2-oxoglutarate is produced from glutamine and glutamate through deamidation, ultimately entering the TCA cycle. Exercise also reduces levels of glutamate and leucine, whose breakdown can further elevate 2-oxoglutarate levels. Furthermore, elevated levels of glutamate are observed in populations with MetS when compared to those without. This supports the idea that exercise boosts CRF, promotes the synthesis of 2-oxoglutarate in the TCA cycle, and helps alleviate the adverse effects of MetS risk factors.

Arginine (Arg) is an amino acid involved in various metabolic pathways. It serves as a substrate for the NOS enzyme family, which produces nitric oxide (NO), a key molecule involved in normal endothelial function, insulin sensitivity, and metabolic profile. Research has demonstrated that during cardiopulmonary exercise testing, the concentration of arginine rises noticeably in heart failure patients, and, in those with greater exercise capacity, the concentration of arginine is positively connected to exercise capacity [24]. This is consistent with the views expressed in this essay. Additionally, Holz et al. [25] showed that 30 min of exercise at a 75% VO_{2max} intensity resulted in differential levels of glutamate and arginine. This suggests that exercise training positively affects endothelial function by significantly enhancing the endothelium-dependent dilatation of the musculoskeletal system, leading to notable changes in arginine levels, which are linked to NO production. Arg supplementation has been shown to considerably improve endothelial [26,27], β-cell [28], and oxidative stress function [29] in patients with T2D, according to current studies. Consistent with Palmnas' findings, this study concluded that the level of arginine in the HCHM group was significantly higher compared to that in the LCLM group [30]. In conclusion, there is evidence supporting the idea that physical activity can enhance health by increasing arginine levels in the body and, to a lesser extent, reducing the severity of illnesses. Exercise could potentially serve as an alternative to arginine supplementation therapy; however, additional research is necessary to confirm this hypothesis.

Serine (Ser) is a neuronal nutrition factor and a precursor of numerous important molecules, including D-serine, phosphatidylserine, sphingomyelin, and glycine. Low levels of Ser have been associated with risk factors such as metabolic syndrome, DM, obesity, and IR. In this study, Ser levels were observed to be higher in subjects with high CRF levels and lower in subjects with high MetS risk factors. Studies have shown that low levels of Ser, glycine, and threonine can differentiate between individuals with poor metabolic profiles (≥2 MetS risk factors) and those with good metabolic profiles (<2 MetS risk factors) [31]. In order to measure energy expenditure and physical activity levels, Lee et al. [13] compared the highest (50%) and lowest (50%) participants in both groups using the doubly labeled water. Ser levels were significantly higher in subjects with HCHM when compared to LCLM. This change may be related to the potential mechanism of the negative correlation between sphingolipid metabolism and Ser. Dube et al. [32] found that both diet-induced weight loss and exercise training can improve IR, but only exercise can reduce the levels of ceramide. Ser may be utilized in obese conditions as a precursor of sphingolipids, potentially leading to ceramide accumulation in insulin-sensitive organs (such as the liver and muscle), thereby accelerating the onset of IR [33]. High levels of exercise with high CRF may downregulate Serine Palmitoyl Transferase (SPT), the enzyme catalyzing the initial step of de novo sphingolipid synthesis by utilizing Ser and palmitoyl coenzyme A to generate 3-ketosphinganine. The lipid-mediated suppression of insulin signaling, and ceramide buildup are both prevented by the inhibition of SPT. Thus, it makes sense to speculate that SPT at high CRF levels would be crucial for ceramide metabolism and IR in individuals.

Cis-aconitic acid, along with its derivatives, is involved in the metabolism of the tricarboxylic acid (TCA) cycle, acetic acid, and dicarboxylic acids. In the TCA cycle, cis-aconitic acid catalyzes the conversion of citrate to isocitrate, thereby playing a crucial role in the completion of the entire TCA cycle process. In this study, we observed that the concentration of cis-aconitic acid was significantly higher in the high-level CRF group when compared to the low-level CRF group. Shi et al. [34] took blood samples from 20 marathon runners both before and after the race. They discovered that the levels of cis-aconitic acid, pyruvic acid, malic acid, glyceric acid, and galacturonic acid were much higher after the race than they were before. This suggests that cis-aconitic acid levels in the body can be considerably raised by aerobic endurance activity. The HM group exhibited significantly lower levels of cis-aconitic acid compared to the LM group. Additionally, Zou et al. [35] observed that women with T2D and severe obesity had lower concentrations of TCA cycle intermediates, including citric acid, cis-aconitic acid, and γ -ketoglutaric acid. These findings suggest that severe obesity and T2D may be associated with an impaired TCA cycle. In this study, the HCHM group exhibited significantly higher levels of cis-aconitic acid compared to the LCLM group. This may be attributed to the fact that the HCHM group engaged in long-term aerobic endurance exercise, where energy is predominantly derived from the aerobic metabolism of fats and carbohydrates. This enhances the activity of the body's TCA cycle and leads to the accumulation of various TCA cycle intermediates, resulting in significant increases in metabolites associated with the metabolism pathways of acetaldehyde and dicarboxylic acid. To explain this phenomenon, individuals with high CRF levels exhibited a greater capacity for the TCA cycle when compared to those with low CRF levels and low MetS risk factors, despite individuals with high MetS risk factors showing lower levels of cis-aconitic acid.

Glutamine constitutes 20% of the total pool of free amino acids in the human body, making it the most abundant amino acid [36]. The liver, lungs, and adipose tissue are the primary sources of glutamine in the human blood as they are capable of synthesizing and releasing it. According to this study, the HC group's glutamine level was substantially lower than the LC group. Several research works have demonstrated that prolonged endurance training or high-intensity training causes blood glutamine levels to drop [37].Although the mechanisms underlying this phenomenon are unknown, they could be due to an increase in glutamine absorption by other tissues, a decrease in muscle-released glutamine, or a decrease in muscle-synthesized glutamine [38]. However, glutamine levels were significantly higher in the HM group than in the LM group, where D-glutamine and D-glutamate metabolic pathways were significantly affected. Alsoud et al. [39] found that the glutamine level in the pre-DM MetS group was 4.8 times higher than that of the control group, and in the normoglycemic MetS group, it was 3.5 times higher. Notably, glutamine can be converted to glutamate, which synthesizes glutathione along with cysteine and glycine, so dysregulated glutathione synthesis may exacerbate the pathogenesis of DM [40]. Glutamate levels in the HCHM group were lower than in the LCLM group. Additionally, Lee et al. [13] observed that after a 12-week intervention combining strength and endurance exercises, concentrations of several metabolites involved in glutathione biosynthesis—including glutamate, total cysteine, total glutathione, creatinine, and taurine—were reduced. This suggests that chronic exercise may decrease glutathione biosynthesis. This effect could be attributed to enhanced insulin sensitivity and improved mitochondrial function, particularly at high levels of CRF.

Valine is one of the BCAAs, which is mostly metabolized in skeletal muscle. Plasma levels of valine were lower in the HC group compared to in the LC group; plasma levels of valine were higher in the HM group compared to in the LM group. Based on recent research, BCAA may be a biomarker for obesity, IR, and T2D. People with greater valine levels had a 36% higher chance of developing T2D compared to those with lower levels [41]. Wientzek et al. [18] found that increasing high-intensity physical activity (>6MET/h) by 1 h per day could reduce BCAA levels by 185% of the standard deviation, possibly due to its association with the TCA cycle. Valine undergoes oxidation and decomposition within the mitochondria of skeletal muscles, generating succinyl coenzyme A to enter the TCA cycle. Given the close relationship between CRF, mitochondrial density, and oxidative enzyme activity in skeletal muscles, it can be inferred that the active TCA cycle promotes the breakdown of valine, thereby reducing its concentration in the blood. This study also found that, when compared to the LCLM group, valine levels decreased in the HCHM group. Studies have demonstrated that aerobic exercise reduces inflammation, enhances BCAA aminotransferase activity, and upregulates the expression of genes related to the TCA cycle [42,43]. These alterations in enzyme activity encourage BCAA breakdown, which lowers plasma levels of the amino acid.

In summary, 2-oxoglutaric acid, arginine, serine, cis-aconitic acid, glutamine, and valineine are not only potential biomarkers reflecting different levels of CRF and different degrees of MetS risk factors, but also demonstrate that high CRF levels are protective factors influencing changes in the metabolic status of MetS risk factors.

4.3. Limitation

The metabolomic analysis of blood from MetS risk factor populations with varying CRF levels was conducted using LC-MS technology to compare metabolic profiles across varying CRF levels and to assess the impact of CRF on different stages of MetS risk factors, elucidating the metabolic mechanisms underlying how changes in CRF levels influence MetS risk factors, challenging existing paradigms regarding the impact of CRF on MetS risk factors. Nevertheless, further investigation is warranted to address the following unresolved issues. The blood metabolic profiles of populations at risk for MetS with varying levels of CRF were comprehensively analyzed using untargeted metabolomics technology to elucidate the impact of CRF on factors related to MetS; future studies could include quantitative analyses using targeted metabolomics technology to further investigate the diagnostic relevance of identified biomarkers. This study exclusively utilized metabolomics techniques due to experimental constraints. Exercise metabolomics research aims to integrate various omics approaches (e.g., genomics, proteomics) to comprehensively elucidate the biological implications of exercise. Therefore, the integration of other omics techniques to provide a comprehensive understanding of the biological effects of exercise.

5. Conclusions

Untargeted metabolomics technology was used to determine the plasma metabolic profiles of MetS risk factor populations with different CRF levels. Based on the comprehensive screening and validation procedures, eight metabolites were identified as markers distinguishing MetS risk factor populations based on CRF levels, namely a-ketoglutarate, cis-aconitic acid, arginine, γ-aminobutyric acid, serine, valine, methionine, and glutamine, indicating significant metabolic profile differences between MetS risk factor populations with high and low CRF levels. Arginine biosynthesis, the TCA cycle, and cysteine and methionine metabolism were markedly influenced by high CRF levels, implying a potential role of high CRF levels in metabolic improvements among individuals with MetS risk factors.

Author Contributions: Conceptualization, J.L. and X.F.; methodology, Q.H.; software, X.F., and Q.H.; visualization, X.F.; formal analysis, X.F.; investigation, X.F.; resources, J.L.; data curation, X.F. and Q.H.; writing—original draft preparation, Q.H., X.F.; writing—review and editing, J.L.; visualization, J.L.; supervision, J.L.; project administration, J.L.; funding acquisition, J.L. All authors have read and agreed to the published version of the manuscript.

Funding: This study was supported by the key projects in the Ministry of education of Humanities and Social Science project grant, Number: 20YJA890014.

Institutional Review Board Statement: The study was conducted in accordance with the Declaration of Helsinki, and approved by the Institutional Review Board (or Ethics Committee) of JIMEIUNIVERSITY OF INSTITUTE (protocol code JMU202207032 and 12 May 2020 of approval).

Informed Consent Statement: Informed consent was obtained from all subjects involved in the study.

Data Availability Statement: Data is contained within the article.

Acknowledgments: We express our gratitude to the subjects for their participation and adherence to experimental protocol in this study and no author is affiliated with companies or manufacturers who will benefit from the results of the present study.

Conflicts of Interest: The authors declare no conflicts of interest.

References

1. Marten, R.; McIntyre, D.; Travassos, C.; Shishkin, S.; Longde, W.; Reddy, S.; Vega, J. An Assessment of Progress towards Universal Health Coverage in Brazil, Russia, India, China, and South Africa (BRICS). *Lancet* **2014**, *384*, 2164–2171. [CrossRef]
2. Bye, A.; Høydal, M.A.; Catalucci, D.; Langaas, M.; Kemi, O.J.; Beisvag, V.; Koch, L.G.; Britton, S L.; Ellingsen, Ø.; Wisløff, U. Gene Expression Profiling of Skeletal Muscle in Exercise-Trained and Sedentary Rats with Inborn High and Low VO_{2max}. *Physiol. Genomics* **2008**, *35*, 213–221. [CrossRef]
3. Martin, S.D.; McGee, S.L. The Role of Mitochondria in the Aetiology of Insulin Resistance and Type 2 Diabetes. *Biochim. Biophys. Acta* **2014**, *1840*, 1303–1312. [CrossRef] [PubMed]
4. Heaney, L.M.; Deighton, K.; Suzuki, T. Non-Targeted Metabolomics in Sport and Exercise Science. *J. Sport Sci.* **2019**, *37*, 959–967. [CrossRef] [PubMed]
5. Wientzek, A.; Floegel, A.; Knüppel, S.; Vigl, M.; Drogan, D.; Adamski, J.; Pischon, T.; Boeing, H. Serum Metabolites Related to Cardiorespiratory Fitness, Physical Activity Energy Expenditure, Sedentary Time and Vigorous Activity. *Int. J. Sport Nutr. Exerc. Metab.* **2014**, *24*, 215–226. [CrossRef]
6. Chorell, E.; Svensson, M.B.; Moritz, T.; Antti, H. Physical Fitness Level Is Reflected by Alterations in the Human Plasma Metabolome. *Mol. BioSyst.* **2012**, *8*, 1187. [CrossRef] [PubMed]
7. Floegel, A.; Wientzek, A.; Bachlechner, U.; Jacobs, S.; Drogan, D.; Prehn, C.; Adamski, J.; Krumsiek, J.; Schulze, M.B.; Pischon, T.; et al. Linking Diet, Physical Activity, Cardiorespiratory Fitness and Obesity to Serum Metabolite Networks: Findings from a Population-Based Study. *Int. J. Obes.* **2014**, *38*, 1388–1396. [CrossRef] [PubMed]
8. Okekunle, A.P.; Li, Y.; Liu, L.; Du, S.; Wu, X.; Chen, Y.; Li, Y.; Qi, J.; Sun, C.; Feng, R. Abnormal Circulating Amino Acid Profiles in Multiple Metabolic Disorders. *Diabetes Res. Clin. Pract.* **2017**, *132*, 45–58. [CrossRef] [PubMed]
9. Kujala, U.M.; Vaara, J.P.; Kainulainen, H.; Vasankari, T.; Vaara, E.; Kyröläinen, H. Associations of Aerobic Fitness and Maximal Muscular Strength with Metabolites in Young Men. *JAMA Netw. Open* **2019**, *2*, e198265. [CrossRef] [PubMed]
10. Morris, C.; Grada, C.O.; Ryan, M.; Roche, H.M.; De Vito, G.; Gibney, M.J.; Gibney, E.R.; Brennan, L. The Relationship between Aerobic Fitness Level and Metabolic Profiles in Healthy Adults. *Mol. Nutr. Food Res.* **2013**, *57*, 1246–1254. [CrossRef]

11. Sardeli, A.V.; Castro, A.; Gadelha, V.B.; Santos, W.M.D.; Lord, J.M.; Cavaglieri, C.R.; Chacon-Mikahil, M.P.T. Metabolomic Response throughout 16 Weeks of Combined Aerobic and Resistance Exercise Training in Older Women with Metabolic Syndrome. *Metabolites* **2022**, *12*, 1041. [CrossRef] [PubMed]
12. Aledo, J.C. Methionine in Proteins: The Cinderella of the Proteinogenic Amino Acids. *Protein Sci.* **2019**, *28*, 1785–1796. [CrossRef] [PubMed]
13. Lee, S.; Olsen, T.; Vinknes, K.J.; Refsum, H.; Gulseth, H.L.; Birkeland, K.I.; Drevon, C.A. Plasma Sulphur-Containing Amino Acids, Physical Exercise and Insulin Sensitivity in Overweight Dysglycemic and Normal Weight Normoglycemic Men. *Nutrients* **2018**, *11*, 10. [CrossRef] [PubMed]
14. Ristow, M.; Zarse, K. How Increased Oxidative Stress Promotes Longevity and Metabolic Health: The Concept of Mitochondrial Hormesis (Mitohormesis). *Exp. Gerontol.* **2010**, *45*, 410–418. [CrossRef]
15. Falone, S.; Mirabilio, A.; Pennelli, A.; Cacchio, M.; Di Baldassarre, A.; Gallina, S.; Passerini, A.; Amicarelli, F. Differential Impact of Acute Bout of Exercise on Redox- and Oxidative Damage-Related Profiles between Untrained Subjects and Amateur Runners. *Physiol. Res.* **2010**, *59*, 953–961. [CrossRef]
16. Gong, L.; Yang, S.; Zhang, W.; Han, F.; Xuan, L.; Lv, Y.; Liu, H.; Liu, L. Targeted Metabolomics for Plasma Amino Acids and Carnitines in Patients with Metabolic Syndrome Using HPLC-MS/MS. *Dis. Markers* **2020**, *2020*, 8842320. [CrossRef] [PubMed]
17. Azab, S.M.; De Souza, R.J.; Lamri, A.; Shanmuganathan, M.; Kroezen, Z.; Schulze, K.M.; Desai, D.; Williams, N.C.; Morrison, K.M.; Atkinson, S.A.; et al. Metabolite Profiles and the Risk of Metabolic Syndrome in Early Childhood: A Case-Control Study. *BMC Med.* **2021**, *19*, 292. [CrossRef]
18. Yan, Y.; Song, G.; Zhu, W.G.; Pan, X.L. A Comparative Study on Blood Metabolomics Characteristics of People with Different Exercise Habits. *J. Beijing Sport Univ.* **2021**, *44*, 34–46. (In Chinese)
19. Tian, J.; Dang, H.N.; Yong, J.; Chui, W.-S.; Dizon, M.P.G.; Yaw, C.K.Y.; Kaufman, D.L. Oral Treatment with γ-Aminobutyric Acid Improves Glucose Tolerance and Insulin Sensitivity by Inhibiting Inflammation in High Fat Diet-Fed Mice. *PLoS ONE* **2011**, *6*, e25338. [CrossRef]
20. Yuan, Y.; Xu, P.; Jiang, Q.; Cai, X.; Wang, T.; Peng, W.; Sun, J.; Zhu, C.; Zhang, C.; Yue, D.; et al. Exercise-induced A-ketoglutaric Acid Stimulates Muscle Hypertrophy and Fat Loss through OXGR1-dependent Adrenal Activation. *EMBO J.* **2021**, *40*, e108434. [CrossRef]
21. Yuan, Y.; Zhu, C.; Wang, Y.; Sun, J.; Feng, L.; Ma, W.; Li, L.; Peng, T.; Yin, C.; Xu, L.; et al. alpha-Ketoglutaric acid ameliorates hyperglycemia in diabetes by inhibiting hepatic gluconeogenesis via serpina1e signaling. *Sci. Adv.* **2022**, *8*, eabn2879. [CrossRef]
22. Liang, W.D.; Huang, P.J.; Xiong, L.H.; Zhou, S.; Ye, R.Y.; Liu, H.; Wei, H.; Lai, R.Y. Metabolomics and its application in the mechanism analysis on diabetic bone metabolic abnormality. *Eur. Rev. Med. Pharmacol. Sci.* **2020**, *24*, 9591–9600. [PubMed]
23. Leibowitz, A.; Klin, Y.; Gruenbaum, B.; Gruenbaum, S.; Kuts, R.; Dubilet, M.; Ohayon, S.; Boyko, M.; Sheiner, E.; Shapira, Y.; et al. Effects of Strong Physical Exercise on Blood Glutamate and Its Metabolite 2-Ketoglutarate Levels in Healthy Volunteers. *Acta Neurobiol. Exp.* **2012**, *72*, 385–396. [CrossRef] [PubMed]
24. Drohomirecka, A.; Waś, J.; Wiligórska, N.; Rywik, T.M.; Komuda, K.; Sokołowska, D.; Lutyńska, A.; Zieliński, T. L-Arginine and Its Derivatives Correlate with Exercise Capacity in Patients with Advanced Heart Failure. *Biomolecules* **2023**, *13*, 423. [CrossRef]
25. Holz, O.; DeLuca, D.S.; Roepcke, S.; Illig, T.; Weinberger, K.M.; Schudt, C.; Hohlfeld, J.M. Smokers with COPD Show a Shift in Energy and Nitrogen Metabolism at Rest and during Exercise. *Int. J. Chron. Obstruct. Pulmon. Dis.* **2020**, *15*, 1–13. [CrossRef] [PubMed]
26. Monti, L.D.; Setola, E.; Lucotti, P.C.G.; Marrocco-Trischitta, M.M.; Comola, M.; Galluccio, E.; Poggi, A.; Mammì, S.; Catapano, A.L.; Comi, G.; et al. Effect of a Long-term Oral L-arginine Supplementation on Glucose Metabolism: A Randomized, Double-blind, Placebo-controlled Trial. *Diabetes Obes. Metab.* **2012**, *14*, 893–900. [CrossRef]
27. Lucotti, P.; Setola, E.; Monti, L.D.; Galluccio, E.; Costa, S.; Sandoli, E.P.; Fermo, I.; Rabaiotti, G.; Gatti, R.; Piatti, P. Beneficial Effects of a Long-Term Oral L-Arginine Treatment Added to a Hypocaloric Diet and Exercise Training Program in Obese, Insulin-Resistant Type 2 Diabetic Patients. *Am. J. Physiol. Endocrinol. Metab.* **2006**, *291*, E906–E912. [CrossRef]
28. Shi, X.Y.; Hou, F.F.; Niu, H.X.; Wang, G.B.; Xie, D.; Guo, Z.J.; Zhou, Z.M.; Yang, F.; Tian, J.W.; Zhang, X. Advanced Oxidation Protein Products Promote Inflammation in Diabetic Kidney through Activation of Renal Nicotinamide Adenine Dinucleotide Phosphate Oxidase. *Endocrinology* **2008**, *149*, 1829–1839. [CrossRef]
29. Vasilijević, A.; Buzadžić, B.; Korać, A.; Petrović, V.; Janković, A.; Korać, B. Beneficial Effects of L-arginine–Nitric Oxide-producing Pathway in Rats Treated with Alloxan. *J. Physiol.* **2007**, *584*, 921–933. [CrossRef]
30. Palmnäs, M.S.A.; Kopciuk, K.A.; Shaykhutdinov, R.A.; Robson, P.J.; Mignault, D.; Rabasa-Lhoret, R.; Vogel, H.J.; Csizmadi, I. Serum Metabolomics of Activity Energy Expenditure and Its Relation to Metabolic Syndrome and Obesity. *Sci. Rep.* **2018**, *8*, 3308. [CrossRef]
31. Batch, B.C.; Shah, S.H.; Newgard, C.B.; Turer, C.B.; Haynes, C.; Bain, J.R.; Muehlbauer, M.; Patel, M.J.; Stevens, R.D.; Appel, L.J.; et al. Branched Chain Amino Acids Are Novel Biomarkers for Discrimination of Metabolic Wellness. *Metabolism* **2013**, *62*, 961–969. [CrossRef] [PubMed]
32. Dubé, J.J.; Amati, F.; Toledo, F.G.S.; Stefanovic-Racic, M.; Rossi, A.; Coen, P.; Goodpaster, B.H. Effects of Weight Loss and Exercise on Insulin Resistance, and Intramyocellular Triacylglycerol, Diacylglycerol and Ceramide. *Diabetologia* **2011**, *54*, 1147–1156. [CrossRef] [PubMed]

33. Samuel, V.T.; Shulman, G.I. Mechanisms for Insulin Resistance: Common Threads and Missing Links. *Cell* **2012**, *148*, 852–871. [CrossRef]
34. Shi, R.; Zhang, J.; Fang, B.; Tian, X.; Feng, Y.; Cheng, Z.; Fu, Z.; Zhang, J.; Wu, J. Runners' Metabolomic Changes Following Marathon. *Nutr. Metab.* **2020**, *17*, 19. [CrossRef] [PubMed]
35. Zou, K.; Turner, K.; Zheng, D.; Hinkley, J.M.; Kugler, B.A.; Hornby, P.J.; Lenhard, J.; Jones, T.E.; Pories, W.J.; Dohm, G.L.; et al. Impaired Glucose Partitioning in Primary Myotubes from Severely Obese Women with Type 2 Diabetes. *Am. J. Physiol. Cell Physiol.* **2020**, *319*, C1011–C1019. [CrossRef] [PubMed]
36. Altman, B.J.; Stine, Z.E.; Dang, C.V. From Krebs to Clinic: Glutamine Metabolism to Cancer Therapy. *Nat. Rev. Cancer* **2016**, *16*, 619–634. [CrossRef]
37. Agostini, F.; Biolo, G. Effect of Physical Activity on Glutamine Metabolism. *Curr. Opin. Clin. Nutr. Metab. Care* **2010**, *13*, 58–64. [CrossRef] [PubMed]
38. Dos Santos, R.V.; Caperuto, É.C.; De Mello, M.T.; Batista, M.L., Jr.; Rosa, L.F. Effect of Exercise on Glutamine Synthesis and Transport in Skeletal Muscle from Rats. *Clin. Exp. Pharma Physiol.* **2009**, *36*, 770–775. [CrossRef]
39. Alsoud, L.O.; Soares, N.C.; Al-Hroub, H.M.; Mousa, M.; Kasabri, V.; Bulatova, N.; Suyagh, M.; Alzoubi, K.H.; El-Huneidi, W.; Abu-Irmaileh, B.; et al. Identification of Insulin Resistance Biomarkers in Metabolic Syndrome Detected by UHPLC-ESI-QTOF-MS. *Metabolites* **2022**, *12*, 508. [CrossRef]
40. Lu, S.C. Glutathione Synthesis. *Biochim. Biophys. Acta* **2013**, *1830*, 3143–3153. [CrossRef]
41. Guasch-Ferré, M.; Hruby, A.; Toledo, E.; Clish, C.B.; Martínez-González, M.A.; Salas-Salvadó, J.; Hu, F.B. Metabolomics in Prediabetes and Diabetes: A Systematic Review and Meta-Analysis. *Diabetes Care* **2016**, *39*, 833–846. [CrossRef] [PubMed]
42. Laferrère, B.; Reilly, D.; Arias, S.; Swerdlow, N.; Gorroochurn, P.; Bawa, B.; Bose, M.; Teixeira, J.; Stevens, R.D.; Wenner, B.R.; et al. Differential Metabolic Impact of Gastric Bypass Surgery versus Dietary Intervention in Obese Diabetic Subjects Despite Identical Weight Loss. *Sci. Transl. Med.* **2011**, *3*, 80re2. [CrossRef] [PubMed]
43. Stump, C.S.; Short, K.R.; Bigelow, M.L.; Schimke, J.M.; Nair, K.S. Effect of Insulin on Human Skeletal Muscle Mitochondrial ATP Production, Protein Synthesis, and mRNA Transcripts. *Proc. Natl. Acad. Sci. USA* **2003**, *100*, 7996–8001. [CrossRef] [PubMed]

metabolites

Article

Carbohydrate and Fat Oxidation in Muscle Assessed with Exercise Calorimetry in 6465 Subjects

Jean-Frédéric Brun [1,*], Emmanuel Varlet [1], Justine Myzia [1], Emmanuelle Varlet-Marie [2], Eric Raynaud de Mauverger [1] and Jacques Mercier [1]

1 Unité d'Explorations Métaboliques et Musculaires (UEMM), Département de Physiologie Clinique, Hôpital Lapeyronie-CHU de Montpellier, Université de Montpellier, 34295 Montpellier, France; emmanuel.varlet@etu.umontpellier.fr (E.V.)

2 UFR des Sciences Pharmaceutiques et Biologiques, Laboratoire du Département de Physicochimie et Biophysique, Institut des Biomolécules Max Mousseron (IBMM), Université de Montpellier, 34093 Montpellier, France; e-varlet_marie@chu-montpellier.fr

* Correspondence: docteurbrunvalescimur@gmail.com

Abstract

Background/Objectives: Exercise calorimetry provides a means to quantify the relative contributions of lipid and carbohydrate (CHO) oxidation across a range of exercise intensities. Although lipid oxidation capacity has been widely studied—particularly in relation to exercise prescription for individuals with obesity—the factors governing CHO oxidation during exercise are less clearly defined. This study therefore aimed to investigate, within a large single-center cohort, not only the established determinants of maximal lipid oxidation (LIPOXmax) but also those influencing CHO oxidation. Methods: Exercise calorimetry was performed in a cohort of 6465 individuals (4561 women and 1904 men; mean age 46.5 years; mean BMI 33.6 kg/m^2). Two principal physiological indices were derived: LIPOXmax, defined as the exercise intensity eliciting maximal rates of fat oxidation, and the carbohydrate cost of the watt (CCW), defined as the slope characterizing the relationship between CHO oxidation and power output. Results: LIPOXmax showed positive associations with lean and muscle mass, and negative associations with fat mass and age, supporting the notion that greater muscle mass enhances the capacity for fat oxidation. Although men demonstrated higher absolute maximal fat oxidation rates, adjustment for body composition revealed that women exhibited relatively higher lipid oxidation (+30%, $p < 0.001$), occurring at a greater percentage of $\dot{V}O_{2max}$ (+9.2%, $p < 0.001$). Furthermore, the carbohydrate cost of the watt was significantly elevated in women (+17.8% compared with men). CCW was positively correlated with BMI, fat mass, and age, and negatively correlated with muscle mass, LIPOXmax, and the crossover point—that is, the exercise intensity at which CHO becomes the predominant substrate. Discussion and Conclusions: Individuals with higher adiposity exhibited a greater reliance on carbohydrate oxidation, whereas leaner individuals preferentially oxidized lipids at comparable exercise intensities. These observations reinforce the reciprocal interplay between lipid and carbohydrate metabolism during exercise and highlight the substantial influence of body composition, age, and sex. Notably, this study provides the first comprehensive characterization of the determinants of CHO oxidation during exercise, identifying sex, age, and adiposity as major contributing factors. This underexplored facet of metabolic flexibility may hold practical relevance in clinical contexts such as obesity or susceptibility to exercise-induced hypoglycemia.

Metabolites 2026, 16, 121

https://doi.org/10.3390/metabo16020121

Keywords: lipid metabolism; skeletal muscle; calorimetry; LIPOXmax; FATmax; training; lipid oxidation; fat; maximal fat oxidation (MFO); fat metabolism; indirect calorimetry; peak fat oxidation; substrate oxidation; carbohydrate oxidation; carbohydrate cost of the watt (CCW)

1. Introduction

The metabolic activity of exercising skeletal muscle can be evaluated indirectly and non-invasively through exercise calorimetry. This method, which relies on the measurement of respiratory gas exchange, enables the quantification of carbohydrate and lipid oxidation rates during physical exertion [1–5]. Exercise calorimetry has contributed to the development of targeted strategies for prescribing physical activity in obesity and metabolic disorders by identifying the exercise intensity that elicits maximal lipid oxidation [6–9]. Endurance training performed at this intensity has been implemented with encouraging outcomes in the context of obesity management [6–9].

Accumulating evidence indicates that lipid oxidation rises progressively with increasing exercise intensity, reaches a peak—commonly termed LIPOXmax or FATmax—at approximately 40–50% of maximal aerobic capacity, and subsequently declines, becoming nearly extinguished at higher intensities referred to as LIPOXzero [8] or FATmin [5], typically above 60% of maximal aerobic capacity. In contrast, carbohydrate oxidation continues to increase beyond this threshold. This asymmetric, bell-shaped lipid oxidation profile exhibits substantial interindividual and interpopulation variability and is modulated by numerous factors, including training status, sex, adiposity, dietary patterns, diabetes, hormonal milieu, and pharmacological treatments, as extensively reviewed elsewhere [10].

Exercise calorimetry also enables the quantification of carbohydrate (CHO) oxidation, although this dimension has received considerably less attention. CHO oxidation increases progressively with rising exercise intensity. At low-to-moderate intensities, this increase is approximately linear and can be characterized by the slope of the relationship between CHO oxidation rate and mechanical power output, termed the carbohydrate cost of the watt (CCW) [11,12]. At maximal and supramaximal intensities, the relationship steepens and adopts an exponential profile [13].

Because exercise calorimetry remains insufficiently implemented in routine practice, most published studies are based on relatively small sample sizes [6–10]. At our center, exercise calorimetry has been systematically performed since 2004 to guide individualized exercise prescription in obese and diabetic patients, as well as in athletes and in specific clinical contexts such as anorexia nervosa. This long-standing clinical and research activity has led to the accumulation of an extensive database spanning more than two decades.

The availability of this dataset offered a unique opportunity to delineate the statistical determinants of lipid and carbohydrate oxidation during exercise. Particular emphasis was placed on carbohydrate oxidation, a dimension that remains comparatively underexplored in the existing literature. Accordingly, the primary objective of this study was to identify the determinants of LIPOXmax and the carbohydrate cost of the watt within a large and heterogeneous population. Secondary objectives included comparing these parameters between men and women and assessing their variation across the age spectrum.

https://doi.org/10.3390/metabo16020121

2. Material and Methods

2.1. Subjects

Exercise calorimetry, following the protocol developed in our center and published in 2001, has been routinely implemented since 1997 [3,6] to guide low-intensity endurance training across a range of clinical and athletic contexts, including obesity, diabetes, and various sports disciplines. Beginning in 2004, all tests successfully completed were systematically entered into an Excel database, while tests terminated prematurely were excluded.

As of 2025, the database comprises 6465 individuals (mean age 46.45 ± 0.58 years; range 8–92.4 years), including 4561 women and 1904 men, with broadly comparable age distributions (women: 45.68 ± 0.68 years; men: 48.29 ± 1.11 years) and similar BMI values (women: 33.77 ± 0.50 kg/m^2; men: 33.11 ± 0.76 kg/m^2).

Key exclusion criteria included acute illness, uncontrolled cardiovascular disease, inability to complete the test, and body weight exceeding 150 kg, corresponding to the upper limit of our cycle ergometer.

Overall, this database encompasses individuals with diverse clinical profiles—including type 2 diabetes, metabolic syndrome, and athletic populations—and thus represents a heterogeneous clinical cohort.

Characteristics of patients are presented on Table 1.

Table 1. Anthropometric characteristics of the subjects included in the study (values ± SD). Anthropometry is highly significantly different between males and females in this large sample of unselected subjects but the overall difference in age and BMI is actually very small.

	All Subjects (n = 6465)	Females (n = 4561)	Males (n = 1904)	Comparison t-Test
Age (years)	46.45 ± 0.58	45.68 ± 0.68	48.29 ± 1.11	$p < 0.001$
Height (cm)	166.35 ± 2.07	162.73 ± 2.41	175.02 ± 4.01	$p < 0.001$
Weight (kg)	92.93 ± 1.16	89.36 ± 1.32	101.47 ± 2.32	$p < 0.001$
BMI (kg/m^2) [a]	33.58 ± 0.42	33.70 ± 0.50	33.11 ± 0.76	$p < 0.001$
Fat free mass (kg)	54.36 ± 0.71	49.62 ± 0.77	66.18 ± 1.62	$p < 0.001$
Muscle mass (kg)	24.30 ± 0.33	21.35 ± 0.34	31.96 ± 0.83	$p < 0.001$
% fat	41.12 ± 0.54	43.91 ± 0.68	34.14 ± 0.83	$p < 0.001$
Waist circumference (cm)	103.29 ± 1.35	100.63 ± 1.56	109.93 ± 2.69	$p < 0.001$
Hip circumference (cm)	113.68 ± 1.49	114.91 ± 1.78	110.59 ± 2.71	$p < 0.001$
$\dot{V}O_{2max}$ ACSM [b] (mL^{-1}.min^{-1}.kg^{-1})	25.43 ± 0.32	23.98 ± 0.36	28.91 ± 0.66	$p < 0.001$

[a] BMI: Body Mass Index. [b] $\dot{V}O_{2max}$ ACSM: predicted $\dot{V}O_{2max}$ by ACSM method.

In the sample, BMI values ranged from 15.1 to 61.5 kg/m^2. The distribution by category was as follows:

- Underweight (BMI < 18.5): 0.1% of women and 0.2% of men.
- Normal weight (BMI 18.5–25): 7.9% of women and 7% of men.
- Overweight (BMI 25–30): 21.7% of women and 25% of men.
- Obesity class I (BMI 30–35): 30.2% of women and 33.7% of men.
- Obesity class II (BMI 35–40): 23.9% of women and 20.9% of men.
- Obesity class III (BMI > 40): 16.1% of women and 13.2% of men.

Overall, 70.2% of women and 67.8% of men fell within the obesity range, while only 7.9% of women and 7% of men had a BMI in the normal range.

https://doi.org/10.3390/metabo16020121

Accordingly, this sample represents a wide range of BMIs but includes a large number of patients suffering from overweight and obesity.

2.2. Body Composition

Body composition was evaluated using a BIACORPUS RX 4000 bioelectrical impedance analyzer (MEDI CAL HealthCare GmbH, Greschbachstr. 6a, 76229 Karlsruhe, Germany), which operates with an alternave 50 kHz current [14,15]. The collected impedance data were processed using BodyComp 8.4 software, which estimates segmental fat and fat-free mass through equations developed from the manufacturer's reference population. Skeletal muscle mass was calculated using the Janssen prediction equation, which relies on hand-to-foot resistance measured at 50 kHz [16]. This equation is as follows:

$$\text{Muscle mass (kg)} = H^2/R_{50} \times 0.401 + \text{gender } [M = 1/F = 0] \times 3.825 + \text{age (years)} \times (-0.0731) + 5.102$$

where H is body height and R_{50} is hand-to-foot resistance measured at 50 kHz.

2.3. Exercise Test

Participants performed the exercise assessment in the morning following a 12 h overnight fast. The protocol consisted of five steady-state stages of six minutes each, initially targeted at 20, 30, 40, 50, and 60% of predicted maximal power (Pmax).

Theoretical predicted Pmax values used to target workloads in each subject were calculated according to the very classical empirical formulae of Wasserman [17] that are implemented in our home-made software for calculation of the balance of substrates. First, predicted weight (PW) is calculated from height using sex- and obesity-specific formulas: for men PW = 0.79 × height (cm) − 60.7, for women PW = 0.79 × height (cm) − 68.2, and for obese subjects PW = 0.65 × height (cm) − 42.8, with the actual weight (BW) compared to PW to determine whether an obesity correction must be applied. PW is then used to estimate maximal oxygen uptake (VO_{2max}), according to the formula: VO_{2max} (mL/min) = [50.72 × PW (kg) − 20.40 × age (years) + 5.61] × (1 if male/0.85 if female).

Finally, $\dot{V}O_{2max}$ is converted to Pmax using a standard mechanical efficiency factor (10 mL $O_2 \approx 1$ watt at maximal effort).

These workloads could be adjusted during the procedure depending on the evolution of the respiratory exchange ratio (RER = $\dot{V}CO_2/\dot{V}O_2$) to ensure that values were recorded both below and above 0.9, the threshold corresponding to the Crossover Point defined later. Tests were conducted on an electromagnetically braked cycle ergometer (Ergoline Bosch 500, Ergoline GmbH, 72475 Bitz, Germany). A standard 12-lead ECG continuously tracked heart rate and cardiac activity throughout the protocol. Ventilatory and metabolic variables were obtained using a computerized breath-by-breath analysis system (COSMED Quark CPET, COSMED Srl, Pomezia, Italy).

Predicted $\dot{V}VO_{2max}$ (ACSM Method)

Because a precise measurement of $\dot{V}VO_2$ peak is not essential for exercise calorimetry, maximal oxygen uptake was estimated after exercise was completed, by extrapolating the individual $\dot{V}VO_2$–heart-rate relationship to the theoretical age-predicted maximal heart rate, following ACSM recommendations ($\dot{V}VO_{2max}$_ACSM) [18].

2.4. Exercise Calorimetry

During each workload, $\dot{V}O_2$ and $\dot{V}CO_2$ (mL/min) were recorded to compute the non-protein RER. Rates of lipid oxidation (Lipox) and carbohydrate oxidation (Glucox) were obtained from gas-exchange data using the classical non-protein respiratory quotient

https://doi.org/10.3390/metabo16020121

method previously described [7,8]. This approach provides substrate-utilization values at each exercise intensity. After curve smoothing, two indices describing the shift from lipid toward carbohydrate metabolism were extracted: the maximal lipid-oxidation point (LIPOXmax) and the Crossover Point (COP).

The crossover point has been defined by the team of George Brooks [13] as the point where CHO represents more than 70% of energy used for muscle activity. It can be easily calculated from the equations of calorimetry presented below that this corresponds to a RER value of ≈0.9.

LIPOXmax corresponds to the workload at which lipid oxidation reaches its peak before declining as carbohydrate use continues to rise.

Following earlier work on long-duration steady-state calorimetry [19,20], Perez-Martin et al. proposed a protocol [21] comprising five submaximal stages of 6 min each—the duration required to obtain stable gas-exchange values. The test is carried out on a cycle ergometer with continuous $\dot{V}O_2/\dot{V}CO_2$ monitoring and ECG surveillance. Workloads are initially set to ~30, 40, 50, and 60% of predicted maximal power but can be adapted according to the RER in order to obtain values below and above 0.9 (the level crossover point), and at least one stage above RER = 1, where fat oxidation essentially drops to zero (LIPOXzero or "FATmin"). Gas-exchange measurements from minutes 5–6 of each stage, assumed to reflect steady state, are used to compute substrate oxidation through standard indirect-calorimetry equations [22–24]:

$$\text{Carbohydrate oxidation (mg/min)} = 4.585\,\dot{V}CO_2 - 3.2255\,\dot{V}O_2 \tag{1}$$

$$\text{Lipid oxidation (mg/min)} = -1.7012\,\dot{V}CO_2 + 1.6946\,\dot{V}O_2 \tag{2}$$

The rationale for using 6 min steps and performing calculations on the values of the 5–6th min is based on a study by McRae and coworkers [22] which indicates that, at this time, the CO_2 production from bicarbonate buffers becomes negligible. There has been controversy about the best duration of the steps, and many authors prefer to use shorter steps [23].

Nevertheless, these various protocols provide quite the same information. They show that the increase in lipid oxidation displays an asymmetrical dome-shaped curve. This curve culminates at the level of MFO at an intensity which is termed in this protocol the LIPOXmax, and then lipid oxidation decreases at higher power intensities. The power intensity where it becomes equal to zero is the point where RER is equal to 1 and is termed the LIPOXzero (or FATmin).

The empirical formula (Equation (2)) that gives the lipid oxidation rate is, as reminded above:

$$\text{Lipid oxidation (mg/min)} = -1.7\,\dot{V}CO_2 + 1.7\,\dot{V}O_2 \tag{3}$$

It is easy to deduce from this formula that the relation between power (P) and oxidation of lipids (Lox) displays an asymmetrical dome-shaped curve of the form:

$$\text{Lox} = \text{A.P}\,(1 - \text{RER}) \tag{4}$$

Derivation of this curve enables us to calculate the power intensity at which lipid oxidation becomes maximal, which is the point where the derivative becomes equal to zero. Therefore, the LIPOXmax calculation is only an application of the equation of lipid oxidation used in calorimetry and is model-independent [8].

The reproducibility of the LIPOXmax has been investigated in several studies and it was found to range between 5.02% and 11.4%, and on average 8.7% [8].

https://doi.org/10.3390/metabo1602012

2.5. Measuring the Kinetics of Carbohydrate Oxidation During Exercise

Although the increase in carbohydrate oxidation above resting levels follows an exponential pattern [13], within the range of intensities typically used during graded exercise calorimetry, it behaves, in practice, as an approximately linear function of power output (see Figure 1). The slope of this relationship—representing the carbohydrate cost required to generate one watt of mechanical power—is referred to as the carbohydrate cost of the watt (CCW) [11,12]. As detailed in Aloulou et al. [11], comparative analyses of various mathematical models demonstrate that a linear function provides an adequate and robust description of the rise in CHO oxidation with increasing power output. Alternative curvilinear models do not yield higher correlation coefficients on average, and variance comparisons reveal no significant improvement in data dispersion; indeed, the linear model exhibits the lowest variance among those tested. Although curvilinear models may also produce high correlation coefficients, the linear formulation remains the simplest and most consistently reliable representation of the relationship between CHO oxidation and power output during exercise. In that study, the mean carbohydrate cost per watt (±SEM) was reported as 0.22 ± 0.001 mg min^{-1} kg^{-1} W^{-1}, with the upper boundary of the first quintile at 0.16 and the lower boundary of the fifth quintile at 0.29 mg min^{-1} kg^{-1} W^{-1} [11].

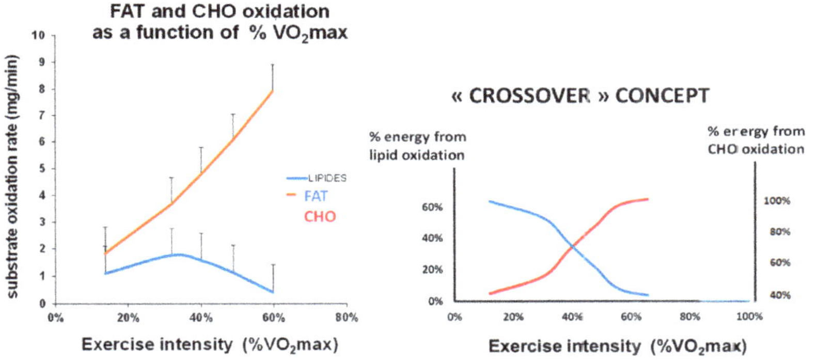

Figure 1. (**left**) Average curve of fat and CHO oxidation as function of exercise intensity in the whole sample of 6465 individuals. It can be seen that over the range of intensities applied during the test CHO oxidation can be approximately modelized as a straight line (carbohydrate cost of the watt) whose slope is on the average 0.24 mg/min/kg/watt. (**right**) Reconstruction from the data of this study of the classical Brook's picture of the "crossover concept" [13].

Intraindividual variability of this parameter was investigated in that previous study. Previous work has shown that the carbohydrate cost of the watt is reproducible with a mean difference of $9.2 \pm 28.4\%$ and a coefficient of variation for paired values of 15.9% [11].

2.6. Statistical Analysis

Data are reported as means ± SD in order to show the distribution of the values, since SEMs are close to 10^{-9}. Statistical analysis was with the Sigmastat package (Version 3.0, Jandel Scientific, Erkrath, Germany). Given the exploratory nature and the very large sample size, *p*-values were interpreted descriptively, with emphasis on effect sizes rather than formal multiplicity correction. Comparisons were performed with two-ways analysis of variance (ANOVA) or Student *t*-test (after the normality of distribution was assessed) when two samples were compared. We validated the assumptions of normality and homogeneity of variances before performing the ANOVA test and calculating the adjusted *p*-values for multiple comparisons using a Tukey test. To account for multiple testing, *p*-values were adjusted for multiple comparisons using Bonferroni's correction.

https://doi.org/10.3390/metabo16020121

Correlations were calculated on Microsoft EXCEL. Differences were considered significant at $p < 0.05$.

3. Results

This large dataset confirms that, although men exhibit higher absolute maximal fat oxidation rates and reach this peak at higher power outputs, women display slightly higher maximal lipid oxidation rates when values are expressed relative to body weight. Moreover, in women, this peak occurs at a marginally higher exercise intensity (Figure 2).

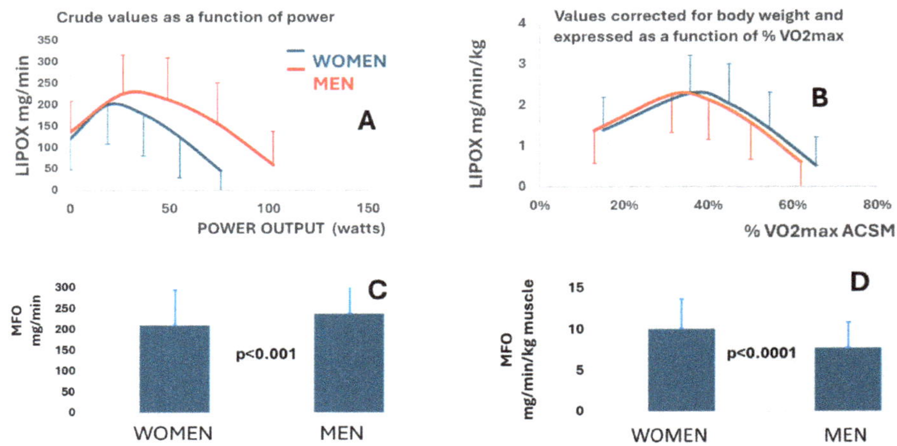

Figure 2. Comparison of men and women for the parameters of lipid oxidation. Error bars represent SD. (**A**) Comparison of fat oxidation expressed as crude values plotted against crude power intensity in men compared to women (ANOVA time effect and group effect $p < 0.001$); (**B**) comparison of fat oxidation corrected for body weight and plotted against % of maximal aerobic capacity (ANOVA time effect and group effect $p < 0.001$); (**C**) comparison of maximal fat oxidation rate expressed as crude values in men compared with women ($p < 0.001$); (**D**) comparison of maximal fat oxidation rate corrected for muscle mass in women compared to men ($p < 0.001$).

More precisely, when lipid oxidation is corrected for anthropometry, women exhibit a higher (+30%) maximal ability to oxidize lipids during exercise, expressed by muscle mass (10.10 ± 0.16 vs. 7.78 ± 0.20 $p < 0.001$) (Table 2), and fat oxidation peaks at a slightly higher (+9.2%) percentage of $\dot{V}O_{2max}$ ACSM (45.62% ± 0.01 vs. 41.73% ± 0.01 $p < 0.001$) (Table 3).

Table 2. Intensity levels of the various flow rates of substrate oxidation during exercise (values ± SD). Anthropometry is highly significantly different between males and females in this large sample of unselected subjects but the overall difference in age and BMI is very small, with average values of the various parameters of balance of substrates during exercise.

	All Subjects (n = 6465)	Females (n = 4561)	Males (n = 1904)	Comparison t-Test
MFO [a] ($mg.min^{-1}$)	218.66 ± 2.72	210.35 ± 3.11	238.54 ± 5.46	$p < 0.001$
MFO [a] corrected for muscle mass ($mg.min^{-1}.kg^{-1}$)	9.46 ± 0.13	10.10 ± 0.16	7.78 ± 0.20	$p < 0.001$
MFO [a] corrected for fat free mass ($mg.min^{-1}.kg^{-1}$)	4.189 ± 1.62	3.63 ± 0.09	4.30 ± 0.07	$p < 0.001$
Crude CCW [b] ($mg.min^{-1}.watt^{-1}$)	21.27 ± 0.26	21.48 ± 0.32	20.75 ± 0.48	$p < 0.001$

https://doi.org/10.3390/metabo16020121

Table 2. *Cont.*

	All Subjects (n = 6465)	Females (n = 4561)	Males (n = 1904)	Comparison t-Test
CCW [b] corrected for body weight ($mg.min^{-1}.kg^{-1}.watt^{-1}$)	0.24 ± 0.004	0.25 ± 0.001	0.21 ± 0.005	$p < 0.001$
CCW [b] corrected for fat-free mass ($mg.min^{-1}.kg^{-1}.watt^{-1}$)	0.41 ± 0.01	0.44 ± 0.01	0.32 ± 0.01	$p < 0.001$
CCW [b] corrected for muscle mass ($mg.min^{-1}.kg^{-1}.watt^{-1}$)	0.94 ± 0.01	1.04 ± 0.02	0.67 ± 0.02	$p < 0.001$

[a] MFO: maximal rate of fat oxidation. [b] CCW: Carbohydrate cost of the watt.

Table 3. Average values of the various parameters of balance of substrates during exercise (values ± SD).

	All Subjects (n = 6465)	Females (n = 4561)	Males (n = 1904)	Comparison t-Test
LIPOX [a] $_{(watt)}$	44.51 ± 0.55	40.56 ± 0.60	53.94 ± 1.23	$p < 0.001$
LIPOX $\dot{V}O_2$	990.18 ± 12.31	930.66 ± 13.78	1132.53 ± 25.93	$p < 0.001$
LIPOXmax% $\dot{V}O_{2max}$	0.44 ± 0.01	45.62 ± 0.01	41.73 ± 0.01	$p < 0.001$
PCX [b] $_{(watt)}$	44.50 ± 0.55	40.55 ± 0.60	53.94 ± 1.23	$p < 0.001$
PCX [b] $\dot{V}O_2$	1038.80 ± 12.91	976.72 ± 14.46	1187.30 ± 27.18	$p < 0.001$
LIPOXzero $\dot{V}O_2$	989.48 ± 12.30	930.54 ± 13.77	1132.55 ± 25.93	$p < 0.001$
LIPOXzero $_{(watt)}$	87.97 ± 1.09	80.06 ± 1.19	106.91 ± 2.45	$p < 0.001$
LIPOXzero% $\dot{V}O_{2max}$	0.67 ± 0.01	0.68 ± 0.01	0.65 ± 0.01	$p < 0.001$

[a] LIPOX: lipid-oxidation point. [b] PCX: crossover point.

The principal correlations between LIPOXmax and other physiological parameters are presented in Figure 3. The power output at which maximal lipid oxidation occurs (LIPOX-max) is positively associated with lean body mass (r = 0.325, $p < 0.001$), a relationship largely driven by its muscular component (r = 0.371, $p < 0.001$). LIPOXmax is also positively correlated with the skeletal muscle index (r = 0.264, $p < 0.001$) and with lean mass expressed as a percentage of total body weight (r = 0.197, $p < 0.001$), indicating that it scales with overall muscularity. Conversely, LIPOXmax is negatively associated with fat mass percentage (r = −0.223, $p < 0.001$) and with age (r = −0.266, $p < 0.001$). A strong negative correlation is also observed with the carbohydrate cost per watt (r = −0.544, $p < 0.001$), reflecting the reciprocal balance between carbohydrate and lipid utilization. Finally, maximal fat oxidation (MFO) and LIPOXmax are themselves positively correlated (r = 0.211, $p < 0.001$).

Carbohydrate oxidation increases progressively with rising power output in both sexes, and within the range of intensities examined, this relationship is approximately linear (Figure 4, upper panel). In women, the rate of increase is 21.48 ± 0.32 mg min^{-1} W^{-1} compared with 20.75 ± 0.48 mg min^{-1} W^{-1} in men (Table 2), resulting in a 17.8% higher carbohydrate cost of the watt in women (0.24 ± 0.004 vs. 0.21 ± 0.005 mg min^{-1} kg^{-1} W^{-1}, $p < 0.001$) (Table 2; Figure 4, lower panel, left). The carbohydrate cost of the watt is also positively associated with age (r = 0.343, $p < 0.001$) (Figure 4, lower panel, right).

https://doi.org/10.3390/metabo16020121

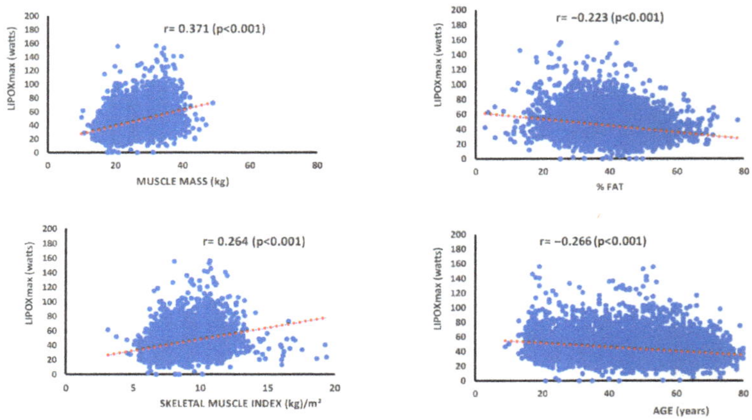

Figure 3. Correlations between parameters quantifying fat oxidation during exercise, age and fatness. **Upper** panel: correlations between the maximal fat oxidation rate at exercise and muscle mass (**left**) and between % fat mass (**right**). **Lower** panel: correlation between LIPOXmax and skeletal muscle index (**left**) and age (**right**).

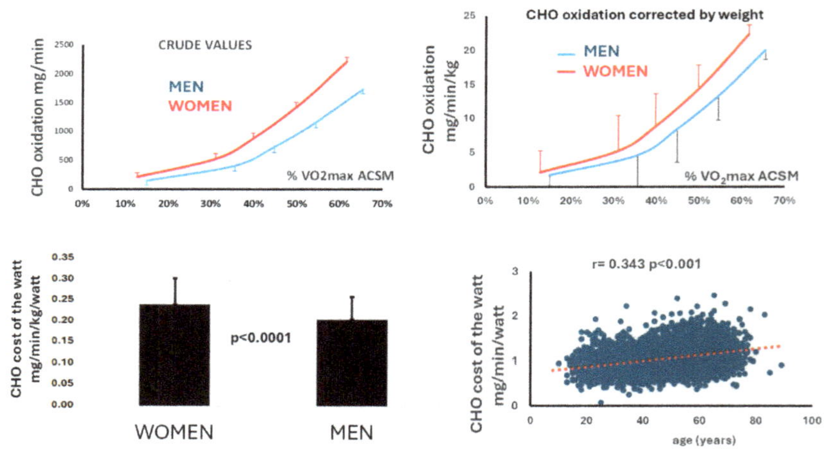

Figure 4. Comparison of men and women for the parameters of Carbohydrate (CHO) oxidation. (**Lower** panel, **right**): carbohydrate cost of the watt is positively correlated with age. (values ± SD).

The carbohydrate cost of the watt expressed as carbohydrate oxidation rate per unit of power (mL/min/watt) is positively correlated with body mass index (r = 0.271 $p < 0.001$) and body fat percentage (r = 0.211 $p < 0.001$), and negatively to the crossover point (r = −0.568 $p < 0.001$) and the LIPOXzero expressed in crude power intensity (r = −0.590 $p < 0.001$) (Figure 5).

When this carbohydrate cost per watt is expressed as carbohydrate oxidation rate per watt and per unit of body weight (mL.min^{-1}.kg^{-1}.watt^{-1}), it is negatively correlated with lean body mass (r = −0.506 $p < 0.001$), muscle mass (r = −0.397 $p < 0.001$), MFO (r = −0.264 $p < 0.001$), the crossover point expressed in crude power (r = −0.546 $p < 0.001$), and LIPOXzero (r = −0.563 $p < 0.001$). The most significant of these correlations are shown in Figure 5.

https://doi.org/10.3390/metabo1602012

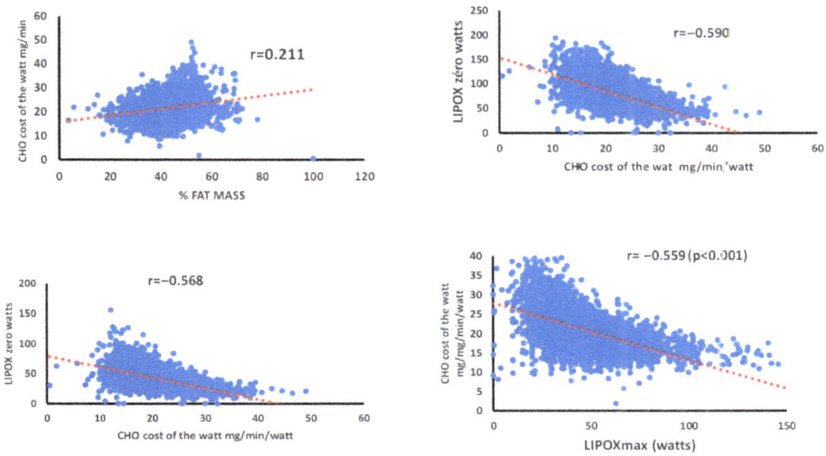

Figure 5. Correlations between parameters quantifying carbohydrate oxidation during exercise, age and fatness. (**Upper** panel, **left**): correlation between the carbohydrate cost of the watt and the percentage of fat mass (r = 0.211 $p < 0.001$); (**Upper** panel, **right**): negative correlation between the carbohydrate cost of the watt and the LIPOX zero, i.e., the intensity level of exercise where CHO becomes the exclusive fuel used for oxidation (r = −0.590 $p < 0.001$). (**Lower** panel, **left**): negative correlation between the carbohydrate cost of the watt and the crossover point expressed as a crude power intensity (r = −0.568 $p < 0.001$). (**Lower** panel, **right**): negative correlation between the carbohydrate cost of the watt corrected for weight and the LIPOXmax expressed as power intensity (r = −0.559 $p < 0.001$).

Stepwise multiple regression analyses were performed to identify the determinants of the substrate-balance parameters described above. Table 4 presents the model for LIPOXmax expressed in absolute power (watts). The model accounted for 15.4% of the variance in the dependent variable (adjusted $R^2 = 0.154$, $p < 0.0001$). All predictors retained in the final model were statistically significant ($p < 0.05$), with age exerting the strongest influence ($\beta = -0.260$), followed by BMI; both variables were negatively associated with LIPOXmax.

Table 4. Final stepwise multiple regression model for dependent variable LIPOXmax expressed in crude power (watts).

Variable	Coefficient (B)	Standard Error	β	F-to-Remove	*p*
Constant	38.905	2.0995			
Age	−0.265	0.0162	−0.260	267.0	<0.0001
BMI	−0.563	0.0910	−0.253	38.2	<0.0001
FFM	0.330	0.0495	0.159	44.4	<0.0001
Muscle mass	0.486	0.0877	0.125	30.7	<0.0001
Fat mass	0.159	0.0429	0.147	13.7	0.0002

Table 5 presents the regression model for LIPOXmax expressed as a percentage of $\dot{V}O_{2max}$. The model accounted for 5.5% of the variance in the dependent variable (adjusted $R^2 = 0.055$, $p < 0.0001$). All predictors retained in the final model were statistically significant ($p < 0.05$), with BMI exerting the strongest influence ($\beta = 0.332$).

https://doi.org/10.3390/metabo16020121

Table 5. Final stepwise multiple regression model for dependent variable LIPOXmax expressed as % VO_{2max}.

Variable	Coefficient (B)	Standard Error	β	F-to-Remove	*p*
Constant	37.9923	2.4169			
Age	0.0688	0.0144	0.0765	22.79	<0.0001
BMI	0.6516	0.0773	0.3323	71.06	<0.0001
FFM	−0.2440	0.0440	−0.1330	30.79	<0.0001
% fat	−0.1413	0.0604	−0.0786	5.47	0.0194

For LIPOXmax expressed either in absolute power or as a percentage of $\dot{V}O_{2max}$, Tables 4 and 5 indicate that the regression models satisfied the assumption of normality but violated homoscedasticity, a limitation that should be taken into account when interpreting the findings. The relatively low adjusted R^2 values suggest that the variables included in the models are, overall, modest predictors of LIPOXmax, and that additional determinants not assessed in this study may also contribute to its variability.

Table 6 presents the regression model for maximal fat oxidation (MFO). The model accounted for 10.8% of the variance in the dependent variable (adjusted R^2 = 0.1077, p < 0.0001). All predictors retained in the final model were statistically significant (p < 0.0001). Muscle mass exerted the strongest influence (β = 0.4809), followed by the skeletal muscle index (SMI; β = −0.3287) and, to a lesser extent, BMI (β = 0.2599).

Table 6. Final stepwise multiple regression model for dependent variable MFO expressed as % VO_{2max}.

Variable	Coefficient (B)	Standard Error	β	F-to-Remove	*p*
Constant	105.99	11.37			
Age	−0.204	0.0861	−0.0386	5.59	<0.0181
BMI	2.986	0.2630	0.2599	128.94	<0.0001
FFM	−0.711	0.3495	−0.0661	4.14	0.0420
Muscle mass	9.706	1.1295	0.4809	73.84	<0.0001
SMI	−19.563	2.706	−0.3287	52.26	<0.0001

The model satisfied the assumption of normality but violated homoscedasticity, a limitation that should be taken into account when interpreting these results.

Table 7 presents the regression model for the carbohydrate cost of the watt (CCW). The model accounted for 36.4% of the variance in the dependent variable (adjusted R^2 = 0.364, p < 0.0001). All predictors retained in the final model were statistically significant (p < 0.0001). LIPOXmax expressed in watts exerted the strongest influence (β = 0.592), followed by fat-free mass (β = −0.415).

Table 7. Final stepwise multiple regression model for dependent variable "Carbohydrate cost of the watt" (CCW).

Variable	Coefficient (B)	Standard Error	β	F-to-Remove	*p*
Constant	4.0488	0.19066			
Age	0.0117	0.00149	0.1096	61.6	<0.0001
FFM	−0.0901	0.00359	−0.4151	631.1	<0.0001
FM	0.0274	0.00169	0.2423	261.1	<0.0001

https://doi.org/10.3390/metabo16020121

Table 7. *Cont.*

Variable	Coefficient (B)	Standard Error	β	F-to-Remove	p
SMI	0.0724	0.01834	0.0603	15.6	<0.0001
LIPOXmax	0.0618	0.00146	0.5916	1783.1	<0.0001

Table 8 shows the regression model for LIPOXzero. This model explained 62.2% of the variance in the dependent variable (adjusted $R^2 = 0.622$, $p < 0.0001$). All predictors included were statistically significant ($p < 0.05$). LIPOXmax expressed as a percentage of $\dot{V}O_{2max}$ had the strongest effect ($\beta = 0.798$), followed by fat mass (kg) ($\beta = -0.166$).

Table 8. Final stepwise multiple regression model for dependent variable "LIPOXzero".

Variable	Coefficient (B)	Standard Error	β	F-to-Remove	p
Constant	0.355075	0.014682			
Age	−0.000373	0.000118	−0.0322	10.05	0.0015
FFM	−0.000886	0.000274	−0.0374	10.48	0.0012
Fat mass	−0.002051	0.000144	−0.1663	203.96	<0.0001
LIPOXmax% $\dot{V}O_{2max}$	0.010308	0.000132	0.7984	6089.46	<0.0001

For both CCW and LIPOXzero, Tables 7 and 8 indicate that the models satisfied the assumption of normality but violated homoscedasticity, a limitation that should be taken into account when interpreting these findings.

4. Discussion

This large dataset offers a unique opportunity to examine the determinants of substrate utilization during exercise across a broad spectrum of ages and body composition profiles. Our findings first confirm the previously reported influence of sex on lipid oxidation. Although men exhibit higher absolute maximal fat oxidation (MFO) values and reach this peak at higher power outputs, anthropometric adjustment reveals that women possess a greater relative capacity to oxidize lipids during exercise. In women, maximal lipid oxidation also occurs at a significantly higher percentage of maximal aerobic capacity (+9.2%). Moreover, the exercise intensity eliciting maximal lipid oxidation is positively associated with fat-free mass—particularly muscle mass—and negatively associated with adiposity. The two principal indices of lipid metabolism, MFO and LIPOXmax, are strongly correlated with each other and both display inverse associations with indices of carbohydrate (CHO) oxidation. Multivariate analyses further indicate that LIPOXmax is negatively related to age and BMI, whereas MFO is primarily determined by muscle mass and, to a lesser extent, BMI. For CCW, the model identifies LIPOXmax as the strongest predictor, followed by fat-free mass, while for LIPOXzero, LIPOXmax is the dominant determinant, followed by fat mass.

Overall, fat oxidation exhibits modest sex-related differences and shows negative associations with age, fat mass, and adiposity. Its magnitude (MFO) is positively correlated with muscle mass. However, these determinants collectively account for only a limited proportion of the variance in fat oxidation.

A particularly notable finding of our study is the central role of the reciprocal balance between lipid and carbohydrate (CHO) utilization during exercise. Indices of CHO oxidation and those of fat oxidation are strongly and inversely correlated, in line with previous work demonstrating a metabolic interaction between these pathways. This pattern

https://doi.org/10.3390/metabo16020121

also reinforces the concept that individuals differ in their preferential substrate use: some predominantly oxidize lipids, whereas others—often described as "glucodependent"—rely more heavily on CHO at a given power output [7,8]. Such profiles are observed not only in sedentary, obese, or diabetic individuals [7,8,21], but also in athletes specializing in short, high-intensity efforts [8].

Overall, the observed correlations do not support the existence of a strict proportional relationship between lipid oxidation and any single determinant. Multivariate analyses produce models that account for only a modest proportion of the variance in LIPOXmax and MFO. Consequently, although statistical prediction models may offer approximate estimates [24], exercise calorimetry remains the most reliable approach for determining individualized training intensities for targeted exercise prescription [6,9].

One of the most distinctive contributions of this study lies in the detailed examination of carbohydrate (CHO) oxidation during exercise, a domain that has received comparatively limited attention in the literature. CHO oxidation increases progressively with rising power output in both sexes, as reflected by the carbohydrate cost of the watt (CCW). This increase is 17.8% higher in women than in men and is positively associated with both age and fat mass. The mean CCW values observed in this cohort closely match those previously reported by Aloulou et al. [11], thereby reinforcing earlier findings. CCW is negatively correlated with LIPOXzero—the exercise intensity at which CHO becomes the exclusive substrate—as well as with the crossover point and LIPOXmax. Overall, CHO oxidation tends to rise with advancing age and increasing adiposity, while decreasing with greater fat-free and muscle mass.

Although maximal lipid oxidation was proposed more than 25 years ago as a tool for exercise prescription in obesity and metabolic diseases, it remains insufficiently implemented in practice. Low-intensity exercise designed to maximize lipid oxidation was long perceived as counterintuitive compared with the higher intensities traditionally associated with athletic training. Yet this intensity domain is readily attainable for sedentary or deconditioned individuals, who often tolerate vigorous exercise poorly. Notably, LIPOXmax closely matches the spontaneous walking pace that can be sustained for prolonged periods without dyspnea [25], a form of locomotion that characterized human activity patterns until the 20th century. Prescribing exercise at LIPOXmax therefore represents a return to a physiologically natural mode of movement that has diminished with modern lifestyles. In contrast, low-volume, high-intensity exercise that primarily stimulates CHO oxidation—such as brief running bouts—has frequently been shown to increase appetite and energy intake. While this effect is offset by the substantial energy expenditure of athletes engaging in high training volumes, individuals performing limited amounts of submaximal high-intensity activity may therefore experience paradoxical and unexpected weight gain [26].

Targeted physical activity performed at LIPOXmax has been extensively evaluated and consistently shown to reduce fat mass while preserving lean and muscle mass and improving aerobic fitness, as highlighted in recent meta-analyses [27,28]. Beyond its excellent tolerability and ease of implementation, this training approach demonstrates notable long-term efficacy, supporting sustained weight stabilization for up to eight years following weight loss [29] and enhancing both the magnitude and durability of the weight-reducing effects of bariatric surgery over a five-year period [30,31].

Consistent with previous reports [32–34], our findings confirm a sex-related difference in lipid oxidation during exercise. This difference is modest and becomes evident only in large cohorts and after adjustment for anthropometric variables and aerobic capacity [35]. Women exhibit a LIPOXmax occurring at an exercise intensity approximately 9% higher than that observed in men. Although the underlying mechanisms remain incompletely

https://doi.org/10.3390/metabo16020121

understood, hormonal influences appear to contribute. Studies by Boisseau and colleagues indicate that estradiol enhances lipid oxidation during exercise, whereas progesterone exerts an opposing effect [36,37]. Additional experimental work further supports the role of estrogens in promoting lipolysis and lipid oxidation in both humans and cultured muscle cells [38–40].

A central objective of this study was to characterize carbohydrate (CHO) oxidation during exercise. We previously proposed quantifying this parameter using the slope of the relationship between CHO oxidation rate and power output (CCW) [11]. Two additional indices complement this approach: the Brooks and Mercier crossover point [13], which is inversely related to CCW and reflects early CHO predominance ("glucodependence") [3], and LIPOXzero, which is negatively correlated with both the crossover point and LIPOXmax. Together, these indices support the concept of a tightly regulated balance between lipid and CHO oxidation, consistent with the crossover model. At the molecular level, this balance is explained by mechanisms described by Sahlin [41], whereby increased intramuscular CHO oxidation inhibits fatty acid entry into mitochondria through CPT-1 inhibition mediated by malonyl-CoA and lactate. Accordingly, the rate of CHO oxidation is one of the key factors governing substrate selection during exercise.

CCW was 17.8% higher in women than in men and increased with both age and fat mass, indicating a greater reliance on CHO oxidation in older and obese individuals. In contrast, higher muscle mass was associated with lower CHO oxidation. In our predominantly sedentary and overweight cohort, this pattern likely reflects concurrent declines in muscle mass and metabolic conditioning. Different profiles might be observed in endurance-trained athletes, who were underrepresented in this sample.

We previously demonstrated that CCW typically remains within narrow limits, behaving almost as a biological constant, although it can be markedly elevated in sedentary individuals or in athletes experiencing exercise-induced hypoglycemia [10–12]. The present study provides robust reference values, confirming those reported earlier in smaller cohorts [11].

Exercise calorimetry has also been proposed as a tool for assessing metabolic flexibility—the capacity of mitochondria to shift between lipid and carbohydrate (CHO) oxidation [10]. Impaired metabolic flexibility is associated with insulin resistance and type 2 diabetes [42] and can be improved through lifestyle interventions [43–45]. The strong inverse relationship observed here between lipid and CHO oxidation supports this concept. In certain situations, however, both lipid oxidation at low intensity and CHO oxidation at high intensity may be simultaneously elevated, as reported in women and in hypothyroid patients treated with L-thyroxine [46]. These observations have contributed to the development of the concept of an optimal fat/CHO oxidation ratio (OLORFOX), which aims to minimize excessive CHO utilization during exercise and has been linked to effects on appetite and weight regulation [47].

Methodologically, substrate balance was assessed using a protocol based on 6 min steady-state exercise stages, originally developed by Perez-Martin et al. [3]. This approach is likely to yield more accurate estimates of substrate oxidation than shorter 3 min stages, which tend to overestimate lipid oxidation and underestimate CHO oxidation in sedentary individuals [48]. Standardization of the test focused on ensuring that measurements captured workloads both above and below a respiratory quotient of 0.9, rather than imposing fixed power outputs, in order to more precisely characterize the substrate crossover zone occurring at this threshold.

This study presents several strengths, most notably its large sample size and the wide range of ages and BMI values represented. However, certain limitations should be acknowledged, including its retrospective, cross-sectional design, the predominance of

https://doi.org/10.3390/metabo16020121

obese participants, and the lack of information on diet, medication use, and training status. Systematic data on diabetes and dyslipidemia were not available for the entire cohort and therefore could not be incorporated as covariates. Nevertheless, the broad BMI range (15.1–61.5 kg/m^2) enhances the generalizability and relevance of the findings.

In this study, confidence intervals around correlation coefficients were not reported. Given the very large sample size, these intervals would be extremely narrow, and emphasis was therefore placed on effect sizes instead. Another methodological consideration concerns the well-documented sex differences in substrate oxidation. Our findings confirm that women and men exhibit slightly different metabolic responses during exercise. The absence of stratified or sex-adjusted analyses may somewhat limit the interpretability of these results. However, our data indicate that these sex-related differences are modest and unlikely to substantially affect the overall conclusions. Nonetheless, a more detailed analysis incorporating sex stratification could provide additional insights and may be worthwhile to pursue.

5. Conclusions

This large-scale study confirms that lipid oxidation during exercise is influenced by sex, age, and adiposity, and provides the first comprehensive characterization of the determinants of CHO oxidation. It underscores the reciprocal balance between the two principal energy substrates, with a shift toward greater "glucodependence" in older and overweight individuals. These findings support the use of exercise calorimetry to assess metabolic flexibility and to refine exercise prescriptions in the contexts of obesity, insulin resistance, and exercise-related hypoglycemia. Preliminary data from this cohort have already contributed to the development of exercise guidelines aimed at optimizing fat oxidation [28], and the present analysis further reinforces their physiological foundation.

Author Contributions: Conceptualization, J.-F.B., E.V.-M. and E.R.d.M.; methodology, J.-F.B., E.V., E.V.-M. and E.R.d.M.; investigation J.-F.B., J.M. (Justine Myzia) and E.V.-M.; writing—original draft preparation, J.-F.B.; writing—review and editing, J.-F.B.; visualization, J.-F.B. and E.V.-M.; supervision, E.R.d.M.; project administration, E.R.d.M. and J.M. (Jacques Mercier). All authors have read and agreed to the published version of the manuscript.

Funding: This research received no external funding.

Institutional Review Board Statement: This study involving human participants was approved by the local Ethics Committee of Montpellier University Hospital (approval code: RB-MTP_2022_03_2022 01039, approval date: 20 March 2022).

Informed Consent Statement: All subjects enrolled in this study received routine hospital care provided by our department. Patients were informed that their clinical results could be used anonymously for research purposes, with the objective of advancing knowledge on metabolism and its regulation, as well as on pathological conditions. Written informed consent was obtained from all participants.

Data Availability Statement: Due to medical confidentiality and ethical restrictions, the patient data underlying this study are not publicly available and cannot be shared.

Conflicts of Interest: The authors declare no conflict of interest.

References

1. Romijn, J.A.; Coyle, E.F.; Sidossis, L.S.; Gastaldelli, A.; Horowitz, J.F.; Endert, E. Regulation of endogenous fat and carbohydrate metabolism in relation to exercise intensity and duration. *Am. J. Physiol.* **1993**, *265*, E380–E391. [CrossRef] [PubMed]
2. Bergman, B.C.; Brooks, G.A. Respiratory gas-exchange ratios during graded exercise in fed and fasted trained and untrained men. *J. Appl. Physiol.* **1999**, *86*, 479–487. [CrossRef]

https://doi.org/10.3390/metabo16020121

3. Perez-Martin, A.; Mercier, J. Stress tests and exercise training program for diabetics—Initial metabolic evaluation. *Ann.* Endocrinol. **2001**, *62*, 291–293.
4. Dériaz, O.; Dumont, M.; Bergeron, N.; Després, J.P.; Brochu, M.; Prud'homme, D. Skeletal muscle low attenuation area and maximal fat oxidation rate during submaximal exercise in male obese individuals. *Int. J. Obes. Relat. Metab. Disord.* **2001**, *25*, 1579–1584. [CrossRef] [PubMed]
5. Achten, J.; Gleeson, M.; Jeukendrup, A.E. Determination of the exercise intensity that elicits maximal fat oxidation. *Med. Sci. Sports Exerc.* **2002**, *34*, 92–97. [CrossRef]
6. Brun, J.F.; Jean, E.; Ghanassia, E.; Flavier, S.; Mercier, J. Metabolic training: New paradigms of exercise training for metabolic diseases with exercise calorimetry targeting individuals. *Ann. Readapt. Med. Phys.* **2007**, *50*, 528–534. [CrossRef]
7. Brun, J.F.; Romain, A.J.; Mercier, J. Maximal lipid oxidation during exercise (Lipoxmax): From physiological measurements to clinical applications. Facts and uncertainties. *Sci. Sports* **2011**, *26*, 57–71. [CrossRef]
8. Brun, J.F.; Varlet-Marie, E.; Romain, A.J.; Mercier, J. Measurement and physiological relevance of the maximal lipid oxidation rate during exercise (LIPOXmax). In *Sports Medicine and Sports Injuries*; INTECH: Hong Kong, China, 2011.
9. Brun, J.F.; Malatesta, D.; Sartorio, A. Maximal lipid oxidation during exercise: A target for individualizing endurance training in obesity and diabetes? *J. Endocrinol. Investig.* **2012**, *35*, 686–691.
10. Brun, J.F.; Myzia, J.; Varlet-Marie, E.; Raynaud de Mauverger, E.; Mercier, J. Beyond the Calorie Paradigm: Taking into Account in Practice the Balance of Fat and Carbohydrate Oxidation during Exercise? *Nutrients* **2022**, *14*, 1605. [CrossRef]
11. Aloulou, I. Le coût glucidique du watt sur ergocycle: Une constante biologique? Carbohydrate cost of the watt on ergocycle: A reproducible biological parameter? *Sci. Sports* **2002**, *17*, 315–317. [CrossRef]
12. Brun, J.F.; Dumortier, M.; Fedou, C.; Mercier, J. Exercise hypoglycemia in nondiabetic subjects. *Diabetes Metab.* **2001**, *27*, 92–106. [PubMed]
13. Brooks, G.A. Importance of the 'crossover' concept in exercise metabolism. *Clin. Exp. Pharmacol. Physiol.* **1997**, *24*, 889–895. [CrossRef] [PubMed]
14. Brun, J.F.; Guiraudou, M.; Mardemootoo, C.; Traoré, A.; Raingeard, I.; Chalançon, A.; Avignon, A. Validation de la mesure segmentaire de la composition corporelle en comparaison avec la DEXA: Intérêt de la mesure de la masse grasse tronculaire. *Sci. Sports* **2013**, *28*, 158–162. [CrossRef]
15. Guiraudou, M.; Maimoun, L.; Dumas, J.M.; Julia, M.; Raingeard, I.; Brun, J.F. Composition corporelle mesurée par impédancemétrie segmentaire (BIAS) et performance de sprint chez les rugbymen. *Sci. Sports* **2015**, *30*, 298–302. [CrossRef]
16. Janssen, I.; Heymsfield, S.B.; Baumgartner, R.N.; Ross, R. Estimation of skeletal muscle mass by bioelectrical impedance analysis. *J. Appl. Physiol.* **2000**, *89*, 465–471. [CrossRef]
17. Wasserman, K.; Hansen, J.; Whipp, B. (Eds.) *Principles of Exercise Testing and Interpretation*; Lea & Febiger: Philadelphia, PA, USA, 1986; pp. 50–80.
18. Aucouturier, J.; Rance, M.; Meyer, M.; Isacco, L.; Thivel, D.; Fellmann, N. Determination of the maximal fat oxidation point in obese children and adolescents: Validity of methods to assess maximal aerobic power. *Eur. J. Appl. Physiol.* **2009**, *105*, 325–331. [CrossRef]
19. Dumortier, M.; Thöni, G.; Brun, J.F.; Mercier, J. Substrate oxidation during exercise: Impact of time interval from the last meal in obese women. *Int. J. Obes.* **2005**, *29*, 966–974. [CrossRef]
20. Manetta, J.; Brun, J.F.; Prefaut, C.; Mercier, J. Substrate oxidation during exercise at moderate and hard intensity in middle-aged and young athletes vs sedentary men. *Metabolism* **2005**, *54*, 1411–1419. [CrossRef]
21. Pérez-Martin, A.; Dumortier, M.; Raynaud, E.; Brun, J.F.; Fédou, C.; Bringer, J.; Mercier, J. Balance of substrate oxidation during submaximal exercise in lean and obese people. *Diabetes Metab.* **2001**, *27*, 466–474.
22. MacRae, H.S.; Noakes, T.D.; Dennis, S.C. Role of decreased carbohydrate oxidation on slower rises in ventilation with increasing exercise intensity after training. *Eur. J. Appl. Physiol. Occup. Physiol.* **1995**, *71*, 523–529. [CrossRef]
23. Jeukendrup, A.E.; Wallis, G.A. Measurement of substrate oxidation during exercise by means of gas exchange measurements. *Int. J. Sports Med.* **2005**, *26*, S28–S37. [CrossRef] [PubMed]
24. Chávez-Guevara, I.A.; Amaro-Gahete, F.J.; Ramos-Jiménez, A.; Brun, J.F. Toward exercise guidelines for optimizing fat oxidation during exercise in obesity: A systematic review and meta-regression. *Sports Med.* **2023**, *53*, 2399–2416. [CrossRef] [PubMed]
25. Dasilva, S.G.; Guidetti, L.; Buzzachera, C.F.; Elsangedy, H.M.; Krinski, K.; De Campos, W.; Goss, F.L.; Baldari, C. Gender-based differences in substrate use during exercise at a self-selected pace. *J. Strength Cond. Res.* **2011**, *25*, 2544–2551. [CrossRef] [PubMed]
26. Brun, J.F.; Romain, A.J.; Sferlazza, A.; Fédou, C.; de Mauverger, E.R.; Mercier, J. Which individuals become fatter when they practice exercise? *Sci. Sports* **2016**, *31*, 214–218. [CrossRef]
27. Romain, A.J.; Carayol, M.; Desplan, M.; Fedou, C.; Ninot, G.; Mercier, J.; Avignon, A.; Brun, J.F. Physical activity targeted at maximal lipid oxidation: A meta-analysis. *J. Nutr. Metab.* **2012**, *2012*, 285395. [CrossRef]

https://doi.org/10.3390/metabo16020121

28. Chávez-Guevara, I.A.; Urquidez-Romero, R.; Pérez-León, J.A.; González-Rodríguez, E.; Moreno-Brito, V.; Ramos-Jiménez, A Chronic Effect of Fatmax Training on Body Weight, Fat Mass, and Cardiorespiratory Fitness in Obese Subjects: A Meta-Analysi of Randomized Clinical Trials. *Int. J. Environ. Res. Public Health* **2020**, *17*, 7888. [CrossRef]
29. Brun, J.F.; Myzia, J.; Varlet-Marie, E.; Mercier, J.; de Mauverger, E.R. The weight-lowering effect of low-intensity endurance trainin targeted at the level of maximal lipid oxidation (LIPOXmax) lasts for more than 8 years, and is associated with improvements i body composition and blood pressure. *Sci. Sports* **2022**, *37*, 603–609. [CrossRef]
30. Brun, J.F.; Lasteyrie, V.; Hammoudi, L.; Nocca, D.; Ghanassia, E. Exercise targeted at the level of maximal lipid oxidatio (LIPOXmax) improves weight loss, decreases orexigenic pulsions and increases satiety after sleeve gastrectomy. *Glob. J. Obe Diabetes Metab. Syndr.* **2019**, *6*, 017–021.
31. Filou, V.; Richou, M.; Bughin, F.; Fédou, C.; de Mauverger, E.R.; Mercier, J.; Brun, J.F. Complementarity of bariatric surgery an physical activity. *Sci. Sports* **2019**, *33*, 65–72. [CrossRef]
32. Friedlander, A.L.; Casazza, G.A.; Horning, M.A.; Huie, M.J.; Piacentini, M.F.; Trimmer, J.K. Training-induced alterations o carbohydrate metabolism in women: Women respond differently from men. *J. Appl. Physiol.* **1998**, *85*, 1175–1186. [CrossRef]
33. Frandsen, J.; Amaro-Gahete, F.J.; Landgrebe, A.; Dela, F.; Ruiz, J.R.; Helge, J.W.; Larsen, S. The influence of age, sex an cardiorespiratory fitness on maximal fat oxidation rate. *Appl. Physiol. Nutr. Metab.* **2021**, *46*, 1241–1247. [CrossRef] [PubMed]
34. Chenevière, X.; Borrani, F.; Sangsue, D.; Gojanovic, B.; Malatesta, D. Gender differences in whole-body fat oxidation kinetic during exercise. *Appl. Physiol. Nutr. Metab.* **2011**, *36*, 88–95. [CrossRef] [PubMed]
35. Brun, J.F.; Boegner, C.; Raynaud, E.; Mercier, J. Contrairement à une idée reçue, les femmes n'oxydent pas plus de lipides à l'effor que les hommes, mais leur LIPOXmax survient à une puissance plus élevée. *Sci. Sports* **2009**, *24*, 45–48. [CrossRef]
36. Boisseau, N.; Isacco, L. Substrate metabolism during exercise: Sexual dimorphism and women's specificities. *Eur. J. Sport Sc* **2022**, *22*, 672–683. [CrossRef]
37. Isacco, L.; Duché, P.; Boisseau, N. Influence of hormonal status on substrate utilization at rest and during exercise in the femal population. *Sports Med.* **2012**, *42*, 327–342. [CrossRef]
38. Mauvais-Jarvis, F.; Clegg, D.J.; Hevener, A.L. The role of estrogens in control of energy balance and glucose homeostasis. *Endoc Rev.* **2013**, *34*, 309–338. [CrossRef] [PubMed] [PubMed Central]
39. Maher, A.C.; Akhtar, M.; Tarnopolsky, M.A. Men supplemented with 17β-estradiol have increased β-oxidation capacity in skeleta muscle. *Physiol. Genom.* **2010**, *42*, 342–347. [CrossRef]
40. Salehzadeh, F.; Rune, A.; Osler, M.; Al-Khalili, L. Testosterone or 17β-estradiol exposure reveals sex-specific effects on glucos and lipid metabolism in human myotubes. *J. Endocrinol.* **2011**, *210*, 219–229. [CrossRef]
41. Sahlin, K.; Sallstedt, E.; Bishop, D.; Tonkonogi, M. Turning down lipid oxidation during heavy exercise—What is the mechanism *J. Physiol. Pharmacol.* **2008**, *59*, 19–30.
42. Kelley, D.E. Skeletal muscle fat oxidation: Timing and flexibility are everything. *J. Clin. Investig.* **2005**, *115*, 1699–1702. [CrossRef
43. San-Millán, I.; Brooks, G.A. Assessment of metabolic flexibility by means of measuring blood lactate, fat, and carbohydrat oxidation responses to exercise in professional endurance athletes and less-fit individuals. *Sports Med.* **2018**, *48*, 467–47 [CrossRef] [PubMed]
44. Benítez-Muñoz, J.A.; Guisado-Cuadrado, I.; Rojo-Tirado, M.Á.; Alcocer-Ayuga, M.; Romero-Parra, N.; Peinado, A.B.; Cupeiro, F Changes in lactate concentration are accompanied by opposite changes in the pattern of fat oxidation: Dose-response relationshi *Eur. J. Sport Sci.* **2024**, *24*, 1653–1663. [CrossRef] [PubMed]
45. Chávez-Guevara, I.A. Assessment of metabolic flexibility by measuring maximal fat oxidation during submaximal intensit exercise: Can we improve the analytical procedures? *Sports Med. Health Sci.* **2023**, *5*, 156–158. [CrossRef] [PubMed]
46. Brun, J.F.; Metrat, S.; Nguyen, J.M.; Richou, M.; Rabta, F.M. Subjects with substituted hypothyroidism oxidize more lipids an carbohydrates during exercise. *Ann. Musculoskelet. Med.* **2018**, *2*, 013–016. [CrossRef]
47. Myzia, J.; Brun, J.F.; Varlet-Marie, E.; de Mauverger, E.R.; Mercier, J. Endurance training minimizing carbohydrate oxidation b targeting the optimal level of fat/carbohydrate oxidation ratio (OLORFOX)? *Sci. Sports* **2022**, *37*, 624–628. [CrossRef]
48. Bordenave, S.; Flavier, S.; Fédou, C.; Brun, J.F.; Mercier, J. Exercise calorimetry in sedentary patients: Procedures based on short min steps underestimate carbohydrate oxidation and overestimate lipid oxidation. *Diabetes Metab.* **2007**, *33*, 379–384. [CrossRef

https://doi.org/10.3390/metabo1602012

metabolites

MDPI

Article

Lifestyle Intervention Therapy Modulates Global DNA Methylation and Adipogenic Gene Expression in Severely Obese Hypogonadal Men

Siresha Bathina [1,2,*,†], Virginia Fuenmayor Lopez [1,2], Mia Prado [1,2], Salina Biene Teo [1,2], Dennis T. Villareal [1,2], Rui Chen [1,2], Clifford Qualls [3] and Reina Armamento-Villareal [1,2,*]

[1] Division of Endocrinology Diabetes and Metabolism, Baylor College of Medicine, Houston, TX 77030, USA
[2] Department of Medicine, Michael E. DeBakey Veterans Affairs (VA) Medical Center, Houston, TX 77030, USA
[3] Office of Research, New Mexico VA Health Care System, Albuquerque, NM 87108, USA
* Correspondence: sbathina@mdanderson.org (S.B.); reina.villareal@bcm.edu (R.A.-V.); Tel.: +1-713-794-1414 (R.A.-V.)
† Current address: Department of Neurosurgery, MD Anderson Cancer Center, Houston, TX 77030, USA.

Abstract

Background/Objectives: Previous studies have suggested that lifestyle intervention (LSI) therapies involving diet and exercise can modulate DNA methylation; however, whether this occurs in severely obese hypogonadal men undergoing weight loss from diet and exercise remains unclear. **Methods**: In this study, we investigated the effects of weight loss from diet and exercise on global DNA methylation as well as on the mRNA expression of specific demethylation enzymes, *DNMT1*, *DNMT3A*, and *DNMT3B*—in peripheral blood mononuclear cells (PBMCs) and DNA methylation markers in DNA of severely obese hypogonadal men. This is a secondary analysis of samples of severely obese (body mass index of $\geq$35 kg/m^2) hypogonadal men undergoing weight loss from diet and exercise in addition to an aromatase inhibitor (anastrozole) or placebo for a total of 12 months. **Results**: LSI therapy significantly reduced global DNA methylation and 5-methylcytosine (5-mC) levels, decreased *DNMT1*, *DNMT3A*, and *DNMT3B* ($p < 0.05$) mRNA levels and markedly decreased *CEBPα*, *FTO*, and *PPARγ* mRNA expression. The reduction in global methylation was independent of aromatase inhibitor use. **Conclusions**: In summary, our findings suggest that LSI induces epigenetic modifications in leukocytes, possibly through the regulation of *DNMT* gene expression. Future studies are warranted to clarify the mechanistic pathways linking lifestyle-induced epigenetic alterations to metabolic health outcomes.

Keywords: obesity; hypogonadism; lifestyle intervention; DNA methylation

1. Introduction

Obesity has become a global epidemic, contributing to a wide spectrum of metabolic and endocrine disorders including type 2 diabetes mellitus, cardiovascular disease, and male hypogonadism [1–3]. The recent literature suggests that obesity and related metabolic dysfunctions are accompanied by alterations in epigenetic regulation, particularly DNA methylation [4], which may mediate the interaction between environmental and genetic factors [5,6].

DNA methylation, the covalent addition of a methyl group to cytosine residues within CpG dinucleotides, is one of the most studied epigenetic modifications regulating gene expression [7]. This process is catalyzed by DNA methyltransferases (*DNMTs*), including *DNMT1*, *DNMT3A*, and *DNMT3B*, which play distinct roles in maintaining and establishing methylation patterns [8]. Aberrant DNA methylation has been implicated in several obesity-related pathologies, influencing genes involved in lipid metabolism, adipogenesis, and insulin sensitivity [9]. Specifically, the peroxisome proliferator-activated receptor gamma (*PPARγ*) and CCAAT/enhancer-binding protein alpha (*CEBPα*) are critical transcription factors governing adipocyte differentiation and glucose homeostasis, and their epigenetic regulation has been associated with obesity and metabolic dysfunction [10,11]. Our group showed that combined diet and exercise resulted in greater improvement in physical function [12] and improvement in metabolic profile [13] than diet or exercise alone. Beyond improving body composition and metabolic health, Keller et al.'s studies on 120 adults (BMI: ~30 kg/m^2) in an 18-month diet +/− physical activity lifestyle randomized trial found that successful weight loss was associated with specific genome-wide DNA methylation changes in blood [14]. Aerobic exercise has been reported to modify DNA methylation in skeletal muscle, adipose tissue, and leukocytes, leading to altered expression of metabolic and inflammatory genes [15]. Dietary factors such as caloric restriction and nutrient composition can also influence methylation status by modulating one-carbon metabolism and methyl donor availability [16]. Despite these insights, the combined effects of diet and exercise on DNA methylation in severely obese hypogonadal men remain largely unexplored.

A prior study in obese individuals found that increased global DNA methylation and elevated *DNMT1* expression have been associated with systemic inflammation and insulin resistance [17]. Conversely, in a rat study, lifestyle modification has been linked to reduced *Dnmt* activity and demethylation of genes regulating metabolism and inflammation [18]. However, most of these findings are derived from pre-clinical and general obese patients, but specific epigenetic mechanisms through which lifestyle interventions (LSIs) influence metabolic outcomes in severely hypogonadal men have not been clearly elucidated. Given these gaps in knowledge, the present study aimed to investigate the effects of LSIs by diet and exercise to promote weight loss on global DNA methylation in peripheral blood mononuclear cells (PBMCs) of obese hypogonadal men. We focused on the methylation and expression of *PPARγ* and *CEBPα* as representative metabolic regulators, and the expression of *DNMT1*, *DNMT3A*, and *DNMT3B* as key enzymes of DNA methylation machinery. By integrating molecular and physiological data, this study seeks to provide mechanistic insights into how LSIs modulate epigenetic pathways which may ultimately contribute to improved metabolic and endocrine function in obese hypogonadal men.

2. Materials and Methods

2.1. Study Design and Patient Population

This was a secondary analysis of DNA and RNA samples obtained from participants of the study, titled "Aromatase inhibitors and weight loss in severely obese men with hypogonadism" (NCT03490513). This study was a randomized double-blind placebo-controlled trial on the effect of weight loss from LSI in combination with an aromatase inhibitor (LSI + AI), anastrozole 1 mg daily, versus placebo (LSI + PBO) over 1 year on hormonal profile and symptoms of hypogonadism in men with severe obesity and hypogonadism. Inclusion/exclusion criteria were as previously published [19], but briefly, the study recruited severely obese men (BMI of $\geq$35 kg/m^2), 35–65 years old, with an average fasting total T done twice between 8 a.m. and 10 a.m. on 2 separate days within 1 month of <300 ng/dL, with luteinizing hormone (LH) of <9.0 mIU/L and estradiol (E2) of $\geq$14 pg/mL, and with symptoms consistent with hypogonadism. The exclusion criteria included the following: (1) clinical/biochemical evidence of hypothalamic/pituitary

https://doi.org/10.3390/metabo16030198

disease; (2) drugs affecting gonadal hormone levels, production and action, or bone metabolism (bisphosphonates, teriparatide, denosumab, glucocorticoids, and phenytoin); (3) diseases affecting bone metabolism (e.g., hyperparathyroidism, untreated hyperthyroidism, osteomalacia, chronic liver disease, significant renal failure, hypercortisolism, malabsorption, immobilization, and Paget's dis.); (4) prostate carcinoma or elevated serum PSA > 4 ng/mL; (5) hematocrit (HCT) more than 50%; (6) untreated severe obstructive sleep apnea; (7) cardiopulmonary disease (e.g., myocardial infarction within 6 months, unstable angina, and stroke) or unstable disease (e.g., NYHA Class III or IV congestive heart failure, severe pulmonary disease requiring steroid pills or the use of supplemental oxygen that would contraindicate exercise or dietary restriction); (8) unstable weight (i.e., ± 2 kg) in the last 3 months; (9) BMD T-score of less than -2.0 at the spine, femoral neck, or total femur; (10) T2DM with fasting blood glucose of >160 mg/dL or A1C of >9.5%.

This study was conducted at Michael E. DeBakey VA Medical Center and the protocol was approved by the Institutional Review Board of Baylor College of Medicine. All participants provided written informed consent in accordance with the guidelines in the Declaration of Helsinki for the ethical treatment of human subjects. This study started in May 2018 and ended in June 2025.

This specific study focused on the methylation changes over time on a subset of subjects with DNA and RNA taken at least at 2 timepoints during the intervention (please see Figure 1 below).

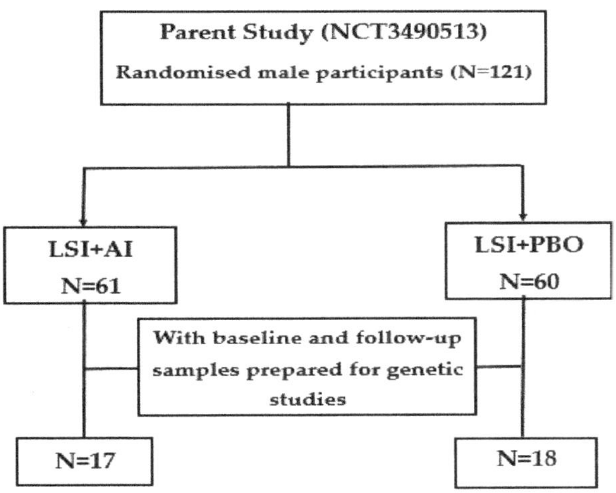

Figure 1. Flow diagram of participants included in the study. LSI + AI—lifestyle intervention + aromatase inhibitor; LSI + PBO—lifestyle intervention+ placebo.

2.2. Diet and Exercise Intervention

Dietary intervention was managed with the aid of a dietitian. Participants were instructed to consume a balanced diet to provide a deficit of 500–750 kcal/day from daily energy requirement. Follow-up visits with dietitian were weekly for the first 3 months and then every 2 weeks thereafter.

Exercise training was supervised by an exercise physiologist and consisted of aerobic training for ~45 min in duration and involved walking on a treadmill and stationary cycling. The resistance training involved nine upper-extremity and lower-extremity exercises with the use of weight-lifting machines. Initially, participants exercised in-person at our facility twice a week. The participants were also instructed to perform home-based exercises involving ground walking, treadmill walking, or stationary biking if available as well as

https://doi.org/10.3390/metabo16030198

resistance exercise using body weight (e.g., abdominal crunches), resistance bands, and ankle weights. However, because of COVID-19 pandemic restrictions, patients were later allowed to perform exercises in a gym of their choice paid for the study or just do home exercises using as much as possible the same routine.

2.3. Body Mass Index (BMI)

Body weight and height were measured by a standard weighing scale and stadiometer, respectively. BMI (kg/m^2) was calculated by dividing the weight (in kilograms) by height (in meters) squared.

2.4. Gene Expression Studies

Blood samples were collected early in the morning after an overnight fast and processed, and then, the samples were stored at −80 °C until analysis. PBMCs were isolated from whole blood using Ficoll density gradient centrifugation. The PBMC fraction contained lymphocytes (T cells, B cells, and NK cells) and monocytes but excluded granulocytes such as neutrophils. Gene expression of *PPARγ*, *CEBPa*, *FTO*, *DNMT1*, *DNMT3A*, and *DNMT3B* in PBMCs was performed by real-time quantitative polymerase chain reaction at baseline (BL) and at 12 months (12 M).

2.4.1. RNA Extraction and qPCR Studies

RNA was extracted from PBMCs using RiboPure Blood (Invitrogen, Carlsbad, CA, USA #AM1928). A total of 200 ng of RNA was used for retro transcription into cDNA and performed using Superscript VILO Master Mix (Invitrogen, Carlsbad, CA, USA) in triplicates following protocol instructions. FAM-labeled TaqMan gene expression assays (Applied Biosystem, College Station, TX, USA) were used for *PPARγ* (Assay ID:), *CEBPα* (Hs00269972_s1:), *FTO* (Hs01057145_m1), *DNMT1* (Hs00945875_m1), *DNMT3A* (Hs01027162_m1), and *DNMT3B* (Hs00171876_m1), and a VIC-labeled TaqMan gene expression assay for housekeeping *18S* (assay ID: Hs03928990_g1). TaqMan Universal Master Mix was used following the manufacturer's protocol. Please see the supplementary details of kits and chemicals used in this study.

Relative quantification: Relative quantification of gene expression of our samples was compared with that of human control total RNA, analyzed by TaqMan-based real-time PCR analysis (Applied Biosystems, #4307281, Carlsbad, CA, USA) was calculated using the DDCT method and adjusted for housekeeping gene expression. Data analysis was performed using a real-time PCR system QuantStudio5 and Quant Studio Design & Analysis Software 1.3.1.

2.4.2. DNA Extraction and Methylation Studies

Genomic DNA was isolated from peripheral leukocytes [20] according to the manufacturer's instructions. DNA quality and quantity were assessed by spectrophotometry (NanoDrop) Bioanalyzer 2100 (Agilent Technologies, Santa Clara, CA, USA). and samples were normalized to 200 ng/μL for methylation analysis. Global DNA methylation (5-methylcytosine, 5-mC) was quantified using the Global DNA Methylation Assay Kit (5-Methyl Cytosine, Colorimetric; Abcam, Cambridge, UK; Cat. No. ab233486) following the manufacturer's instructions.

Global DNA methylation was expressed as the percentage of 5-methylcytosine (5-mC%) relative to total cytosine content and calculated according to the manufacturer's instructions. Values are expressed as the fraction of total methylated cytosine. Because the primary objective of this secondary analysis was to evaluate sustained epigenetic changes following long-term lifestyle intervention, we analyzed specimens collected at baseline

https://doi.org/10.3390/metabo16030198

(BL), 6 months, and 12 months (12 M), representing the full duration of the intervention for global methylation studies.

2.4.3. Statistical Analysis

Overall longitudinal analysis of 5 mC% or global DNA methylation (GM) were performed by repeated measures ANOVA with visit (baseline, 6 months, and 12 months) as the repeated factor adjusted for the baseline outcome value and treatment group (lifestyle with AI or placebo); change in testosterone, estradiol, or weight; or metformin use as covariates. Comparisons of 6- and 12-month values with baseline values for the different methylation parameters were performed by Student's t-test. Data were presented as the means ± SD in tables and text, and the means ± SE in the figures. Data in graphs were analyzed using Prism 9.0 (GraphPad, San Diego, CA, USA) and tables were managed using Excel 2013 (Microsoft, Redmond, WA, USA), analyzed by Statgraphics Centurion XVI X64 (Statgraphics Technologies, Inc., The Plains, VA, USA) and confirmed using SAS version 9.3 (SAS Institute, Inc., Cary, NC, USA) A p-value of < 0.05 was considered statistically significant.

3. Results

Thirty-five participants with RNA and DNA samples taken at least at 2 timepoints were included in this study (Figure 1). The average baseline weight of these subjects was 135.9 ± 20.5 kg, and the baseline BMI was 43.8 ± 5.3 kg/m^2. Table 1 below shows the baseline characteristics of the participants included in the study. There were no significant differences in clinical characteristics between those randomized to LSI + AI compared to the LSI + PBO at baseline. In addition, there were no significant differences in baseline characteristics of the entire population and the subjects included in this study (see Supplementary Table S2).

Table 1. Baseline characteristics of the subjects.

Body Parameter	Anastrozole (n = 17)	Placebo (n = 18)	p-Value
Age	54.1 ± 5.2	49.8 ± 8.0	0.07
BMI	42.3 ± 5.9	44.4 ± 4.5	0.26
Weight	131.0 ± 22.4	140.8 ± 16.0	0.15
Racial background			
White	7	10	0.30
Blacks	10	7	
Type 2 diabetes	11	7	0.17
Metformin	7	7	0.85
Hemoglobin A1c	6.9 ± 1.1	6.8 ± 1.4	0.92
Testosterone	221.5 ± 47.8	235.1 ± 47.5	0.41
Estradiol	26.3 ± 17.9	21.4 ± 5.0	0.29

Values are means ± SD, BMI-body mass index.

3.1. Body Composition

At the end of the study, the subjects lost an average of −3.7 ±3.6% (−6.7 ± 7.1 kg). Testosterone significantly increased in the LSI + AI group compared to the LSI + PBO group (+174.6+/−134.3 ng/dL vs. +43.8+/−103.5 ng/dL; respectively, $p = 0.01$), and estradiol significantly decreased in the LSI + AI group compared to LSI + PBO group (−13.9+/−19.0 pg/mL vs. +1.5+/−7.7 pg/mL; respectively, $p = 0.01$) at the end of 12 months. Table 2 below shows the

https://doi.org/10.3390/metabo16030198

changes in body composition of the participants compared to baseline at 6 and 12 months. Body fat, visceral adipose tissue, and lean mass (total and appendicular) decreased from baseline. However, these changes were not statistically significant.

Table 2. Changes (%) in body composition.

Body Parameter	6 Months	12 Months
Total body fat %	−2.2 ± 4.9	−1.7 ± 7.2
Total body fat (g)	−5.1 ± 7.8	−4.9 ± 11.0
Visceral adipose tissue	−8.0 ± 25.5	−3.8 ± 32.7
Total lean mass	−1.2 ± 3.3	−1.7 ± 3.8
Appendicular lean mass	−1.1 ± 8.0	−4.8 ± 6.0

The above table illustrates % changes in body composition. Data are presented as mean ± SD. Measurements were obtained at baseline and follow-up visits (6 and 12 months). *p*-values are not significant compared to baseline by Student's *t*-test.

3.2. DNMTs Expression Is Downregulated Following Diet and Exercise Intervention

Lifestyle with or without aromatase inhibitors (LSI ± AI) significantly modulated the expression of DNA methylation-related genes (Figure 2). Reduction in mRNA expression was observed at 12 months compared to baseline for *DNMT1* (BL: 3.29 ± 1.1 vs. 12 M: 0.78 ± 0.16, $p = 0.038$), *DNMT3A* (BL: 1.28 ± 0.18 vs. 12 M: 0.81 ± 0.09, $p = 0.028$), and *DNMT3B* (BL: 0.97 ± 0.15 vs. 12 M: 0.64 ± 0.07, $p = 0.05$) following the intervention (Figure 2a–c). Collectively, these findings demonstrate a consistent downregulation of key components of DNA methylation machinery after the 12-month lifestyle program.

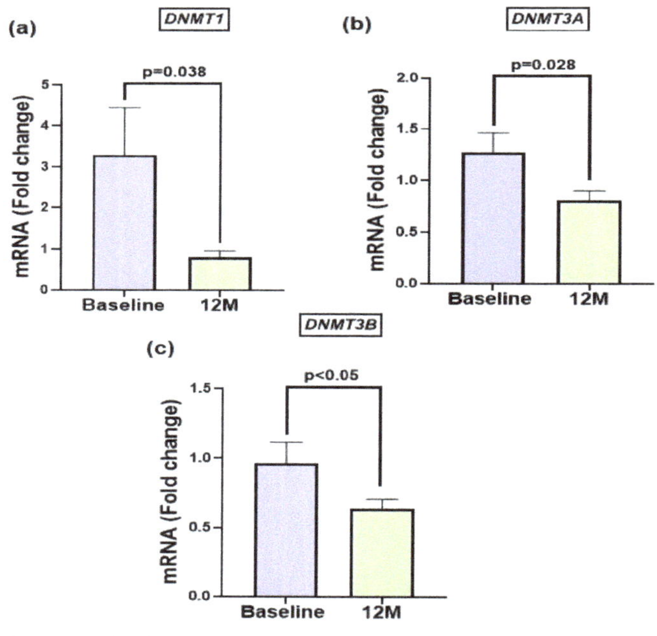

Figure 2. Lifestyle ± aromatase inhibitors downregulate DNA methyltransferase (*DNMT*) gene expression in PBMCs. mRNA expression significantly decreased after 12 months of lifestyle intervention compared with baseline for *DNMT1* ($p = 0.038$) (**a**), *DNMT3A* ($p = 0.028$) (**b**), and *DNMT3B* ($p < 0.05$) (**c**). Data are presented as mean fold change ± SEM.

https://doi.org/10.3390/metabo16030198

3.3. Lifestyle Modification Reduces Global and 5-mC% DNA Methylation

To validate whether the transcriptional changes were accompanied by epigenetic alterations at the DNA level, global DNA methylation markers were assessed in the PBMC samples collected at baseline (BL), 6 months (6 M), and 12 months (12 M). LSI ± AI significantly reduced the percentage of 5-mC%, with levels decreasing from BL (0.18 ± 0.03) to 6 M (0.103 ± 0.005; $p = 0.052$) and 12 M (0.081 ± 0.004; $p = 0.013$) (Figure 3a). Similarly, global DNA methylation showed a downward trajectory, declining from BL (0.85 ± 0.18) to 6 M (0.49 ± 0.03; $p = 0.053$) and 12 M (0.39 ± 0.014; $p = 0.015$) (Figure 3b). Collectively, these data indicate that LSI ± AI promotes gradual genomic DNA demethylation in PBMCs over the 12-month intervention period.

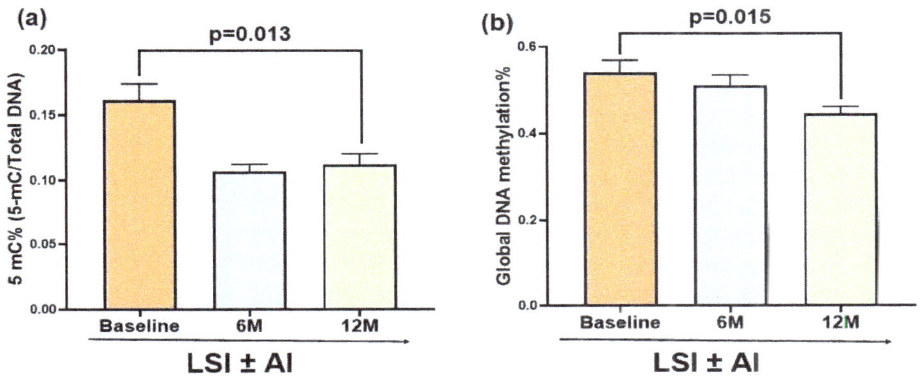

Figure 3. Lifestyle intervention with or without an aromatase inhibitor (LSI ± AI) reduces global DNA methylation and 5-methylcytosine % (5-mC%) levels in DNA of buffy coat samples. There was a decrease from baseline in 5-mC% levels at both follow-up timepoints but significant only at 12 months ($p = 0.013$) (**a**) and global DNA methylation percentage in PBMCs across the same timepoints which was significant only at 12 M ($p = 0.015$) with intervention (**b**). Bars represent the mean ± SEM.

To examine the influence of AI, as half of the subjects on this study were on it and hormone studies showed significant differences in testosterone and estradiol levels between those who were and were not on AI, we compared the change in 5-mC% and global methylation at 6 and 12 months between the two groups. We found significant differences over time (6 months and 12 months; $p = 0.002$ for 5-mC% and $p < 0.001$ for global methylation by repeated measures ANOVA) but no differences due to treatment with AI or PBO ($p = 0.12$ for 5-mC% and $p = 0.06$ for global methylation), testosterone ($p = 0.62$ for 5-mC% and $p = 0.21$ for global methylation), or estradiol ($p = 0.35$ for 5-mC% and $p = 0.92$ for global methylation); and an independent difference due to weight ($p = 0.02$ for 5-mC% and $p = 0.03$ for global methylation). Since some of our patients were on metformin which has been known to affect methylation, we adjusted for the intake of the drug which did not show any difference between those on LSI + AI and LSI + PBO (Table 3).

Table 3. Changes in methylation with lifestyle intervention ± aromatase inhibitors.

	LSI + AI	LSI + PBO	*p*	* *p*	** *p*
Weight change (kg)					
6 months	−5.8 ± 6.8	−3.8 ± 6.9	0.44		
12 months	−7.1 ± 7.8	−6.2 ± 6.3	0.72		

https://doi.org/10.3390/metabo16030198

Table 3. *Cont.*

	LSI + AI	LSI + PBO	*p*	* *p*	** *p*
5-mC%					
6 months	−0.32 ± 0.39	−0.29 ± 0.55	0.86	0.71	
12 months	−0.17 ± 0.43	−0.08 ± 0.59	0.62	0.47	0.002
%Global methylation					
6 months	−0.05 ± 0.08	−0.01 ± 0.22	0.67	0.79	
12 months	−0.17 ± 0.16	−0.04 ± 0.21	0.12	0.12	<0.001

p adj—* *p* adjusted for the intake of metformin and ** *p* adjusted by repeated measures ANOVA across visits (6 and 12 months). Note: there was no difference due to treatment with an aromatase inhibitor or placebo ($p = 0.12$ for 5-mC% and $p = 0.06$ for global methylation), testosterone ($p = 0.62$ for 5-MC% and $p = 0.21$ for global methylation), or estradiol ($p = 0.35$ for 5-mC% and $p = 0.92$ for global methylation), and an independent difference due to weight ($p = 0.02$ for 5-mC% and $p = 0.03$ for global methylation). LSI + AI—lifestyle intervention + aromatase inhibitor; LSI + PBO—lifestyle intervention + placebo.

3.4. Lifestyle Intervention Alters Expression of Adipogenic and Metabolic Genes

LSI ± AI significantly modulated the expression of genes associated with adipogenesis and metabolism in PBMCs in men with severe obesity and hypogonadism. LSI ± AI also decreased mRNA levels of adipogenic markers (Figure 4), *PPARγ* (BL: 1.53 ± 0.37 vs. 12 M: 0.83 ± 0.18; $p = 0.08$), and *CEBPα* (BL: 1.59 ± 0.32 vs. 12 M: 0.76 ± 0.18; $p = 0.045$). Similarly, *FTO* mRNA expression was significantly decreased after 12 months of LSI ± AI (BL: 1.47 ± 0.27 vs. 12 M: 0.86 ± 0.09; $p = 0.028$) (Figure 4). Notably, the direction of change was consistent across participants, and the coordinated downregulation of *PPARγ*, *CEBPα*, and *FTO* suggests a broad transcriptional response to the LSI ± AI.

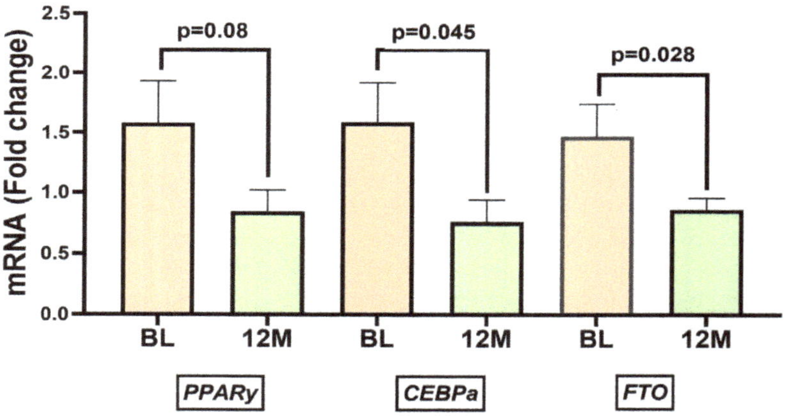

Figure 4. Lifestyle intervention with or with aromatase inhibitors (LSI ± AI) alters expression of adipogenic and metabolic genes. mRNA expression levels of *PPARγ*, *CEBPα*, and *FTO* were measured in PBMCs before (BL/baseline) and after the lifestyle intervention (12 M). Bars represent the mean fold change ± SEM. Expression of all three genes significantly decreased following the intervention, indicating reduced adipogenic and metabolic activity associated with LSI ± AI.

4. Discussion

This study provides novel evidence that combined diet and exercise elicit significant epigenetic and transcriptional adaptations in severely obese hypogonadal men. We observed reductions in global DNA methylation, decreased expression of DNA methyltransferases (*DNMT1*, *DNMT3A*, and *DNMT3B*), and downregulation of key metabolic

https://doi.org/10.3390/metabo1603019

genes (*PPARγ*, *CEBPα*, and *FTO*) following 12 months of LSI ± AI. Our analysis also indicated that treatment with AI or PBO showed no influence on the changes in methylation. Thus, our findings suggest that the combination of diet and exercise reprograms the leukocyte methylome, which likely accompanies the metabolic improvement associated with LSI.

DNA methylation, catalyzed by DNMTs, regulates gene expression, genomic stability, and imprinting [7]. Hyperactivation of DNMTs has been linked to obesity, insulin resistance, and chronic inflammation [17]. The human genome encodes five *DNMTs*—*DNMT1*, *DNMT2*, *DNMT3A*, *DNMT3B*, and *DNMT3L*. Among these, *DNMT1*, *DNMT3A*, and *DNMT3B* are canonical cytosine-5 methyltransferases which contain an N-terminal regulatory region and a C-terminal catalytic domain that uses S-adenosylmethionine (SAM) as a methyl donor and a base-flipping mechanism to generate 5-methylcytosine [21], representing global methylation. *DNMT1* preferentially methylates hemi-methylated DNA and is therefore considered the principal maintenance methyltransferase [22], whereas *DNMT3A* and *DNMT3B* primarily act on unmethylated DNA to establish de novo methylation; dysregulation of these enzymes affects adipose biology [23].

A prior study by Yang et al. suggested adipocyte differentiation involves large epigenomic changes (histone modifications, DNA methylation) in precursor cells (e.g., 3T3-L1) entering adipogenesis in lean murine adipocytes [24]. While other studies showed elevated *DNMT3A* expression in adipose tissue of transgenic mice [23]; these findings underscore the close association between DNMT activity and adipose tissue biology. Although not significantly different from baseline, our subjects experienced a reduction in body fat (total and visceral) with weight loss. The observed reduction in *DNMT1*, *DNMT3A*, and *DNMT3B* expression (Figure 2) after weight loss from diet and exercise suggests that LSI can downregulate methylation machinery, promoting a more transcriptionally permissive chromatin state. These findings are consistent with earlier studies showing exercise-induced hypomethylation [25] and decreased DNMT expression which may also involve other tissues and cells such as skeletal muscle and leukocytes [15].

Although human intervention studies have not yet directly demonstrated suppression of *DNMT1*, *DNMT3A*, or *DNMT3B* in metabolic tissues following lifestyle or exercise interventions, mechanistic evidence from animal and cellular models shows that reduced *DNMT1* can drive passive demethylation [26], while decreased DNMT3A/3B limits de novo methylation [27]. Exercise-regulated DNMT dynamics in muscle and metabolic tissues have been observed in preclinical models [28], supporting our findings. Such changes can activate transcription of genes involved in mitochondrial biogenesis, oxidative metabolism, and anti-inflammatory responses—pathways central to the metabolic improvements associated with sustained physical activity [29].

Despite the emergence of effective weight loss drugs, exercise and dietary modifications remain important components in the management of obesity [30]. LSI not only improves body composition and glucose homeostasis but also remodels the epigenome [31]. Our data indicate that both global 5-mC and total DNA methylation levels were significantly reduced after LSI over 12 months, which was demonstrated only in the PBMCs in our study as we have no other tissues available. These changes occurred independent of AI use. Our observations agree with the results from the studies of Ronn et al., who analyzed subcutaneous adipose tissue biopsies from 23 previously sedentary, healthy men before and after 6 months of endurance exercise training. In this study, the authors observed widespread genome-wide methylation changes, with 17,975 CpG sites (across 7663 genes) significantly altered after training. Moreover, several loci with methylation changes also showed corresponding mRNA expression shifts, including key obesity and type 2 diabetes-related genes TCF7L2 and KCNQ1, demonstrating coordinated epigenetic

https://doi.org/10.3390/metabo16030198

and transcriptional remodeling in metabolic tissues [5]. Nitert et al. also showed the impact of an exercise intervention on genome-wide DNA methylation in human skeletal muscle of patients with diabetes [32].

Based on our findings, we hypothesize that the observed suppression of *DNMT1*, the primary maintenance methyltransferase, may promote passive loss of DNA methylation during cell turnover and concurrent reductions in *DNMT3A* and *DNMT3B* activity, enzymes responsible for de novo methylation, thus further limiting methylation as illustrated in Figure 5. Results from the genome-wide methylation studies of Benton et al. in human adipose tissue also showed differential methylation of obesity-associated genes including *DNMT3A* in obese women before and after gastric bypass [33].

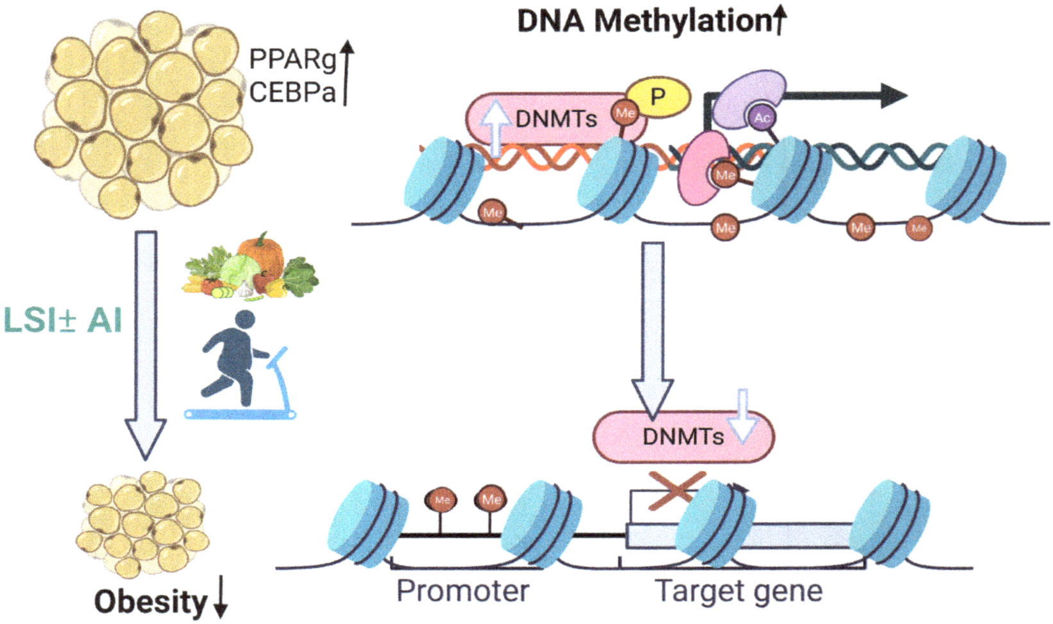

Figure 5. Conceptual mechanism of LSI-induced epigenetic remodeling. The above figure illustrates the proposed mechanism by which lifestyle intervention (LSI) with or without aromatase inhibitors (±AIs) modulates DNA methylation and adipogenic gene regulation. Increased DNA methylation mediated by DNA methyltransferases (*DNMTs*) is shown at regulatory genomic regions, accompanied by histone modifications such as methylation (Me) and acetylation (Ac), leading to altered chromatin structure and transcriptional activity. LSI reduces *DNMT* activity; AI has no independent effect on methylation in this study. This epigenetic remodeling is associated with reduced adipogenesis, reflected by decreased adipocyte accumulation and downregulation of key adipogenic transcription factors, ultimately contributing to reduced obesity.

Weight-loss interventions and reductions in inflammation were often accompanied by partial "reversal" of methylation changes, suggesting a link between inflammation driven by adiposity and epigenetic regulation [34]. In our study, we observed transcriptional downregulation of *PPARγ*, *CEBPα*, and *FTO*, highlighting the metabolic reprogramming induced by LSI. *PPARγ* and *CEBPα* are key adipogenic transcription factors that promote lipid accumulation and adipocyte differentiation [35] This suppression following 12 months of LSI indicates reduced adipogenic signaling which may enhance metabolic efficiency. Similar findings have been reported in exercise interventions, where *PPARγ* promoter demethylation correlated with metabolic adaptations in skeletal muscle [25]. The decline

https://doi.org/10.3390/metabo16030198

in FTO mRNA suggests normalization of the obesity-linked RNA demethylase pathway. FTO regulates energy expenditure and appetite through m6A RNA demethylation; its overexpression is associated with obesity and insulin resistance [36]. Downregulation of FTO following lifestyle modification may contribute to improved metabolic control and energy balance.

5. Limitations

This study is limited to using PBMCs as a surrogate for systemic epigenetic profiling. Although peripheral methylation patterns often reflect whole-body metabolic adaptations, tissue-specific analysis (e.g., skeletal muscle or adipose tissue) would provide stronger mechanistic evidence. However, these tissues are not easily obtainable in humans. This is the reason why some epigenetic studies [37,38] have relied on easily accessible surrogate tissues such as PBMCs. Accordingly, cellular heterogeneity within PBMC samples represents a limitation, and the observed methylation changes should be interpreted with caution and considered primarily as signals warranting further investigation using approaches that control for immune cell composition (e.g., cell sorting), rather than definitive conclusion. Nevertheless, our data provide supporting evidence for reduced methylation activity, as we observed decreased expression of *DNMT*s. In addition, this study included only severely obese hypogonadal men; therefore, our findings may not be generalizable to women, eugonadal or less obese populations. Moreover, our cohort size was small and not powered for subgroup analysis; thus, future studies should include larger populations to confirm differential responses. In addition, a locus-specific methylation analysis would have strengthened our data but was not performed in our study. Finally, half of our subjects were on aromatase inhibitors. However, there was no difference in methylation between those on aromatase inhibitors and those on placebo, suggesting that the changes in methylation patterns in the epigenome of our subjects are primarily due to LSI with contribution from the accompanying weight change.

6. Conclusions

In summary, weight loss from combined lifestyle therapy induces favorable epigenetic reprogramming in obese hypogonadal men, characterized by reduced global DNA methylation, downregulation of *DNMT*s, and suppression of adipogenic genes. Given the small number of subjects in our study, a study with a larger sample size, a longitudinal analysis beyond 12 months, and assessment of locus-specific methylation to assess the stability of these methylation changes and their relationship to long-term clinical outcomes is necessary to confirm our findings. Furthermore, future work integrating genome-wide methylation, transcriptomic, and metabolomic analyses could unravel specific pathways linking methylation to sustainable metabolic improvement.

Supplementary Materials: The following supporting information can be downloaded at: https://www.mdpi.com/article/10.3390/metabo16030198/s1, Table S1: Key resources table; Table S2: Baseline characteristics of the subjects included vs. those not included.

Author Contributions: Conceptualization, S.B. and R.A.-V.; formal analysis, S.B., C.Q. and R.A.-V.; investigation: S.B., V.F.L., M.P., S.B.T., D.T.V., R.C., C.Q. and R.A.-V.; writing of the original draft: S.B.; writing, S.B. and R.A.-V.; reviewing and editing, S.B., V.F.L., M.P., S.B.T., D.T.V., R.C., C.Q. and R.A.-V. All authors have read and agreed to the published version of the manuscript.

Funding: This study was supported by NIH R01 HD093047, VA Merit Review, 101CX001665 awarded to R.A.-V.

Institutional Review Board Statement: The study was conducted at Michael E. DeBakey VA Medical Center (MEDVAMC) in accordance with the guidelines in the Declaration of Helsinki for the ethical

https://doi.org/10.3390/metabo16030198

treatment of human subjects from May 2018 to June 2025. The protocol was approved by the Baylor College of Medicine Institutional Review Board (Approval code: H-41814, Approval date: 11 June 2017) and has Clinical Trials.gov Identifier: NCT03490513.

Informed Consent Statement: All patients provided written informed consent before the beginning of the study.

Data Availability Statement: The original contributions presented in this study are included in the article/Supplementary Materials. Further inquiries can be directed to the corresponding author(s).

Acknowledgments: Preliminary data from this study were submitted and presented in the form of an abstract by S.B at ENDO-2025, the annual scientific meeting of the Endocrine Society [39]. We thank the participants for their cooperation The findings reported in this article are the results of work supported with resources and the use of facilities at Michael E. De Bakey VA Medical Center and the Center for Translational Research in Inflammatory Diseases (CITRID).

Conflicts of Interest: The authors declare no conflicts of interest.

References

1. Bhupathiraju, S.N.; Hu, F.B. Epidemiology of Obesity and Diabetes and Their Cardiovascular Complications. *Circ. Res.* **2016**, *118*, 1723–1735. [CrossRef]
2. Yuen, M.M.A. Health Complications of Obesity: 224 Obesity-Associated Comorbidities from a Mechanistic Perspective. *Gastroenterol. Clin. N. Am.* **2023**, *52*, 363–380. [CrossRef]
3. Ahmed, F.; Hetty, S.; Laterveer, R.; Surucu, E.B.; Mathioudaki, A.; Hornbrinck, E.; Patsoukaki, V.; Olausson, J.; Sundbom, M.; Svensson, M.K.; et al. Altered Expression of Aromatase and Estrogen Receptors in Adipose Tissue from Men with Obesity or Type 2 Diabetes. *J. Clin. Endocrinol. Metab.* **2025**, *110*, e3410–e3424. [CrossRef] [PubMed]
4. Sandovici, I.; Morais, T.; Constância, M.; Monteiro, M.P. Epigenetic Changes Associated with Obesity-related Metabolic Comorbidities. *J. Endocr. Soc.* **2025**, *9*, bvaf129. [CrossRef] [PubMed]
5. Rönn, T.; Volkov, P.; Davegårdh, C.; Dayeh, T.; Hall, E.; Olsson, A.H.; Eriksson, K.F.; Groop, L.; Ling, C. A six months exercise intervention influences the genome-wide DNA methylation pattern in human adipose tissue. *PLoS Genet.* **2013**, *9*, e1003572. [CrossRef]
6. Ling, C.; Rönn, T. Epigenetics in Human Obesity and Type 2 Diabetes. *Cell Metab.* **2019**, *29*, 1028–1044. [CrossRef]
7. Moore, L.D.; Le, T.; Fan, G. DNA Methylation and Its Basic Function. *Neuropsychopharmacology* **2013**, *38*, 23–38. [CrossRef]
8. Lyko, F. The DNA methyltransferase family: A versatile toolkit for epigenetic regulation. *Nat. Rev. Genet.* **2018**, *19*, 81–92. [CrossRef] [PubMed]
9. Barres, R.; Zierath, J.R. DNA methylation in metabolic disorders. *Am. J. Clin. Nutr.* **2011**, *93*, 897s–900s. [CrossRef]
10. Siersbæk, R.; Nielsen, R.; Mandrup, S. Transcriptional networks and chromatin remodeling controlling adipogenesis. *Trends Endocrinol. Metab.* **2012**, *23*, 56–64. [CrossRef]
11. Ge, K. Epigenetic regulation of adipogenesis by histone methylation. *Biochim. Biophys. Acta* **2012**, *1819*, 727–732. [CrossRef] [PubMed]
12. Villareal, D.T.; Chode, S.; Parimi, N.; Sinacore, D.R.; Hilton, T.; Armamento-Villareal, R.; Holbert, N.B.; Brown, L.M.; Mathurin, J.C.; Klein, S. Weight loss, exercise, or both and physical function in obese older adults. *N. Engl. J. Med.* **2011**, *364*, 1218–1229. [CrossRef] [PubMed]
13. Bouchonville, M.; Armamento-Villareal, R.; Shah, K.; Napoli, N.; Sinacore, D.R.; Qualls, C.; Villareal, D.T. Weight loss, exercise or both and cardiometabolic risk factors in obese older adults: Results of a randomized controlled trial. *Int. J. Obes.* **2014**, *38*, 423–431. [CrossRef]
14. Keller, M.; Yaskolka Meir, A.; Bernhart, S.H.; Gepner, Y.; Shelef, I.; Schwarzfuchs, D.; Cohen, N.; Tsaban, G.; Rinott, E.; Zelicha, H.; et al. DNA methylation signature in blood mirrors successful weight-loss during lifestyle interventions: The CENTRAL trial. *Genome Med.* **2020**, *12*, 97. [CrossRef] [PubMed]
15. Barrès, R.; Yan, J.; Egan, B.; Treebak, J.T.; Rasmussen, M.; Fritz, T.; Holloway, C.J.; Mercken, E.M.; Wojtaszewski, J.F.P.; Febbraio, M.A.; et al. Acute exercise remodels promoter methylation in human skeletal muscle. *Cell Metab.* **2012**, *15*, 405–411. [CrossRef] [PubMed]
16. Choi, S.W.; Friso, S. Epigenetics: A New Bridge between Nutrition and Health. *Adv. Nutr.* **2010**, *1*, 8–16. [CrossRef]
17. Kim, A.Y.; Park, Y.J.; Pan, X.; Shin, K.C.; Kwak, S.H.; Bassas, A.F.; Choi, S.H.; Park, Y.M.; Lee, M.K.; Kim, J.B. Obesity-induced DNA hypermethylation of the adiponectin gene mediates insulin resistance. *Nat. Commun.* **2015**, *6*, 7585. [CrossRef]

https://doi.org/10.3390/metabo16030198

18. Elsner, V.R.; Lovatel, G.A.; Moysés, F.; Bertoldi, K.; Spindler, C.; Cechinel, L.R.; Silveira, B.B.; da Silva, P.P.; Horta, A.L.S.; Moreira, J.C.F. Exercise induces age-dependent changes on epigenetic parameters in rat hippocampus: A preliminary study. *Exp. Gerontol.* **2013**, *48*, 136–139. [CrossRef]
19. Bathina, S.; Lopez, V.F.; Prado, M.; Ballato, E.; Colleluori, G.; Tetlay, M.; Villareal, D.T.; Mediwala, S.; Chen, R.; Qualls, C.; et al. Health implications of racial differences in serum growth differentiation factor levels among men with obesity. *Physiol. Rep.* **2024**, *12*, e70124. [CrossRef]
20. Armamento-Villareal, R.; Shah, V.O.; Aguirre, L.E.; Meisner, A.L.; Qualls, C.; Royce, M.E. The rs4646 and rs12592697 polymorphisms in CYP19A1 are associated with disease progression among patients with breast cancer from different racial/ethnic backgrounds. *Front. Genet.* **2016**, *7*, 211. [CrossRef]
21. Tajima, S.; Suetake, I.; Takeshita, K.; Nakagawa, A.; Kimura, H. Domain Structure of the Dnmt1, Dnmt3a, and Dnmt3b DNA Methyltransferases. *Adv. Exp. Med. Biol.* **2016**, *945*, 63–86.
22. Londoño Gentile, T.; Lu, C.; Lodato, P.M.; Tse, S.; Olejniczak, S.H.; Witze, E.S.; Oldham, K.E.; Rothstein, D.; Hock, A.; Ilkayeva, O.R.; et al. DNMT1 is regulated by ATP-citrate lyase and maintains methylation patterns during adipocyte differentiation. *Mol. Cell. Biol.* **2013**, *33*, 3864–3878. [CrossRef] [PubMed]
23. Kamei, Y.; Suganami, T.; Ehara, T.; Kanai, S.; Hayashi, K.; Yamamoto, Y.; Ishii, T.; Shimura, H.; Fujitani, Y.; Kishimoto, T.; et al. Increased expression of DNA methyltransferase 3a in obese adipose tissue: Studies with transgenic mice. *Obesity* **2010**, *18*, 314–321. [CrossRef]
24. Yang, X.; Wu, R.; Shan, W.; Yu, L.; Xue, B.; Shi, H. DNA Methylation Biphasically Regulates 3T3-L1 Preadipocyte Differentiation. *Mol. Endocrinol.* **2016**, *30*, 677–687. [CrossRef] [PubMed]
25. Denham, J.; Marques, F.Z.; Bruns, E.L.; O'Brien, B.J.; Charchar, F.J. Epigenetic changes in leukocytes after 8 weeks of resistance exercise training. *Eur. J. Appl. Physiol.* **2016**, *116*, 1245–1253. [CrossRef] [PubMed]
26. Anderson, R.M.; Bosch, J.A.; Goll, M.G.; Hesselson, D.; Dong, P.D.; Shin, D.; Shimizu, M.; Mourki, J.P.; Mermer, J.E.; Ko, M.S.H.; et al. Loss of Dnmt1 catalytic activity reveals multiple roles for DNA methylation during pancreas development and regeneration. *Dev. Biol.* **2009**, *334*, 213–223. [CrossRef]
27. Okano, M.; Bell, D.W.; Haber, D.A.; Li, E. DNA methyltransferases Dnmt3a and Dnmt3b are essential for de novo methylation and mammalian development. *Cell* **1999**, *99*, 247–257. [CrossRef]
28. Damal Villivalam, S.; Ebert, S.M.; Lim, H.W.; Kim, J.; You, D.; Jung, B.C.; Song, M.; Kim, J.B. A necessary role of DNMT3A in endurance exercise by suppressing ALDH1L1-mediated oxidative stress. *EMBO J.* **2021**, *40*, e106491. [CrossRef]
29. Krammer, U.D.B.; Sommer, A.; Tschida, S.; Mayer, A.; Lilja, S.V.; Switzeny, O.J.; Lackner, J.R.; Krammer, M.B. PGC-1α Methylation, miR-23a, and miR-30e Expression as Biomarkers for Exercise- and Diet-Induced Mitochondrial Biogenesis in Capillary Blood from Healthy Individuals: A Single-Arm Intervention. *Sports* **2022**, *10*, 73. [CrossRef]
30. Kim, H.J.; Kwon, O. Nutrition and exercise: Cornerstones of health with emphasis on obesity and type 2 diabetes management-A narrative review. *Obes. Rev.* **2024**, *25*, e13762. [CrossRef]
31. Ambelu, T.; Teferi, G. The impact of exercise modalities on blood glucose, blood pressure and body composition in patients with type 2 diabetes mellitus. *BMC Sports Sci. Med. Rehabil.* **2023**, *15*, 153. [CrossRef]
32. Nitert, M.D.; Dayeh, T.; Volkov, P.; Elgzyri, T.; Hall, E.; Nilsson, E.; Rönn, T.; Ling, C. Impact of an exercise intervention on DNA methylation in skeletal muscle from first-degree relatives of patients with type 2 diabetes. *Diabetes* **2012**, *61*, 3322–3332. [CrossRef]
33. Benton, M.C.; Johnstone, A.; Eccles, D.; Harmon, B.; Hayes, M.T.; Lea, R.A.; Carr, L.K.; Heckard, D.W.; Luo, J.; Pasquale, L.A.; et al. An analysis of DNA methylation in human adipose tissue reveals differential modification of obesity genes before and after gastric bypass and weight loss. *Genome Biol.* **2015**, *16*, 8. [CrossRef] [PubMed]
34. Fraszczyk, E.; Luijten, M.; Spijkerman, A.M.W.; Snieder, H.; Wackers, P.F.K.; Bloks, V.W.; van der Geijn, M.L.P.; Asselbergs, F.W.; Groen, A.K.; Slagboom, P.E. The effects of bariatric surgery on clinical profile, DNA methylation, and ageing in severely obese patients. *Clin. Epigenetics* **2020**, *12*, 14. [CrossRef]
35. Rosen, E.D.; Hsu, C.H.; Wang, X.; Sakai, S.; Freeman, M.W.; Gonzalez, F.J.; Spiegelman, B.M. C/EBPalpha induces adipogenesis through PPARgamma: A unified pathway. *Genes Dev.* **2002**, *16*, 22–26. [CrossRef]
36. Liu, S.J.; Cai, T.H.; Fang, C.L.; Lin, S.Z.; Yang, W.Q.; Wei, Y.; Liu, X.L.; Zhuang, F.Y.; Zhang, J.J. Long-term exercise training down-regulates m6A RNA demethylase FTO expression in the hippocampus and hypothalamus: An effective intervention for epigenetic modification. *BMC Neurosci.* **2022**, *23*, 54. [CrossRef] [PubMed]
37. Jiang, R.; Jones, M.J.; Chen, E.; Neumann, S.M.; Fraser, H.B.; Miller, G.E.; Kobor, M.S. Discordance of DNA methylation variance between two accessible human tissues. *Sci. Rep.* **2015**, *5*, 8257. [CrossRef]

https://doi.org/10.3390/metabo16030198

38. Perera, F.; Tang, W.Y.; Herbstman, J.; Tang, D.; Levin, L.; Miller, R.; Ho, S.M. Relation of DNA methylation of 5′-CpG island of ACSL3 to transplacental exposure to airborne polycyclic aromatic hydrocarbons and childhood asthma. *PLoS ONE* **2009**, *4*, e4488. [CrossRef]
39. Bathina, S.; Lopez, V.F.; Prado, M.; Teo, S.B.; Davila, V.S.; Chen, R.P.; Villareal, D.T.; Villareal, R.C. MON-707 Lifestyle Intervention Therapy Modulates Global and Gene Specific DNA Methylation in Severely Obese Hypogonadal Men. *J. Endocr. Soc.* **2025**, *9*, bvaf149-044. [CrossRef]

https://doi.org/10.3390/metabo16030198

metabolites

Article

The Metabolome Characteristics of Aerobic Endurance Development in Adolescent Male Rowers Using Polarized and Threshold Model: An Original Research

Fanming Kong [1,2,3,†], Miaomiao Zhu [4,†], Xinliang Pan [5], Li Zhao [6,*], Sanjun Yang [1,3], Jinyuan Zhuo [7], Cheng Peng [2], Dongkai Li [1] and Jing Mi [2,*]

1 Sports Teaching and Research Department, China University of Mining and Technology-Beijing, Beijing 100083, China; 202413@cumtb.edu.cn (F.K.)
2 Sport Coaching College, Beijing Sport University, Beijing 100084, China
3 Institute for Emergency Rescue Ergonomics and Protection, China University of Mining and Technology-Beijing, Beijing 100083, China
4 Liaocheng No. 1 Experimental School, Liaocheng 252001, China; bsuzhumiaomiao@126.com
5 School of Physical Education, Xi'an University of Architecture and Technology, Xi'an 710055, China
6 Sport Science School, Beijing Sport University, Beijing 100084, China
7 Physical Education Department, Renmin University of China, Beijing 100872, China; zhuojinyuan@ruc.edu.cn
* Correspondence: zhaolispring@bsu.edu.cn (L.Z.); taishanmijing@126.com (J.M.)
† These authors contributed equally to this work and should be considered co-first authors.

Abstract: Objective: This study aimed to explore the molecular response mechanisms of differential blood metabolites before and after 8 weeks of threshold and polarized training models using metabolomics technology combined with changes in athletic performance. Methods: Twenty-four male rowers aged 14–16 were randomly divided into a THR group and a POL group (12 participants each). The THR group followed a threshold training model (72%, 24%, and 4% of training time in low-, moderate-, and high-intensity zones, respectively), while the POL group followed a polarized training model (78%, 8%, and 14% training-intensity distribution). Both groups underwent an 8-week training program. Aerobic endurance changes were assessed using a 2 km maximal rowing performance test, and untargeted metabolome analysis was conducted to examine blood metabolomic changes before and after the different training interventions. Aerobic endurance changes were assessed through a 2 km maximal rowing test. Non-targeted metabolomics analysis was employed to evaluate changes in blood metabolome profiles before and after the different training interventions. Results: After 8 weeks of training, both the THR and POL groups exhibited significant improvements in 2 km maximal rowing performance ($p < 0.05$), with no significant differences between the groups. The THR and POL groups had 46 shared differential metabolites before and after the intervention, primarily enriched in sphingolipid metabolism, glutathione metabolism, and glycine, serine, and threonine metabolism pathways. Nine unique differential metabolites were identified in the THR group, mainly enriched in pyruvate metabolism, glycine, serine, and threonine metabolism, glutathione metabolism, and sphingolipid metabolism. A total of 14 unique differential metabolites were identified in the POL group, predominantly enriched in sphingolipid metabolism, glycine, serine, and threonine metabolism, aminoacyl-tRNA biosynthesis, and glutathione metabolism. Conclusions: The 8-week THR and POL training models demonstrated similar effects on enhancing aerobic performance in adolescent male rowers, indicating that both training modalities share similar blood metabolic mechanisms for improving aerobic endurance. Furthermore, both the THR group and the POL group exhibited numerous shared metabolites and some differential metabolites, suggesting that the two endurance training models share common pathways but also have distinct aspects in enhancing aerobic endurance.

 https://doi.org/10.3390/metabo15010017

Keywords: aerobic endurance; polarized training model; threshold training model; rowing; metabolomics

1. Introduction

Metabolomics, as an effective tool for the quantitative and qualitative analysis of endogenous metabolites, operates downstream of gene regulatory and protein interaction networks, providing terminal biological insights. This offers a new perspective for investigating the intrinsic mechanisms of physical exercise [1,2]. Metabolomics refers to the systematic study of all metabolites within a biological sample using mass spectrometry or nuclear magnetic resonance spectroscopy, with a focus on interactions among various systems within the organism [3–5]. It allows for the screening of differential metabolites between individuals and populations while also elucidating the biological mechanisms underlying these differences. With these advantages, metabolomics has emerged as a research hot spot for explaining metabolic changes induced by endurance training [6,7].

Endurance athletes often adopt different intensity distribution models for training, with the threshold training model (THR) and polarized training model (POL) being classic and commonly employed approaches in the field of endurance training. These models are well-supported by theoretical research and have been widely recognized in practical training contexts [8,9]. The POL model's training intensity distribution (TID) focuses primarily on low-intensity (zone 1, Z1) and high-intensity (zone 3, Z3) sessions, whereas the THR model's TID emphasizes Z1 and moderate-intensity sessions (zone 2, Z2). Despite the high proportion of Z1 training in both models, the distinct ratios of Z2 and Z3 lead to different energy metabolism pathways and varying stress responses in the cardiovascular system and skeletal muscles, which may result in differences in blood metabolic pathways [10].

Existing studies on training intensity distribution have largely focused on performance outcomes such as time trials, peak power output, and average power in sports like cross-country skiing [11] and cycling [12]. However, metabolomics studies evaluating and monitoring endurance athletes' physiological states through energy metabolism stages, material transport pathways, and signal transduction mechanisms remain limited [6]. Additionally, factors such as athletic proficiency, training phases, and load measurement standards introduce subjective and objective variations, leaving the effectiveness of these two endurance training models a topic of ongoing debate within academia [13–16]. Given this context, this study selected adolescent male rowers as experimental subjects and applied metabolomics technology combined with changes in athletic performance to explore the molecular mechanisms underlying the development of aerobic endurance before and after 8 weeks of THR and POL training. This aims to provide scientific evidence for enhancing the efficiency and quality of endurance training in adolescent athletes.

2. Materials and Methods

2.1. Participants and Group

This study involved 24 adolescent male rowers from a sports school in Shandong Province as experimental subjects, who were randomly divided into the THR group and the POL group, with 12 participants in each group. During the study period, all participants ate and lived at the sports school under consistent conditions, with strict restrictions on the use of mobile phones and other electronic devices. The intake of protein supplements, nutritional supplements, and sports drinks was also prohibited. Detailed information on the participants is provided in Table 1.

Table 1. Basic characteristics of the research subjects (n = 24).

Group	n	Bone Age (Years)	Height (cm)	Weight (kg)	BMI (kg/m^2)	Body Fat Percentage (%)	Training Years (Years)
THR	12	15.38 ± 1.21	1.79 ± 0.06	71.13 ± 11.76	22.12 ± 2.51	11.52 ± 4.17	1.36 ± 0.61
POL	12	15.68 ± 1.11	1.80 ± 0.05	72.23 ± 7.42	22.39 ± 2.05	13.29 ± 3.55	1.35 ± 0.78

Note: Data are presented as mean ± standard deviation. THR = threshold training model; POL = polarized training model; BMI = Body Mass Index.

2.2. Study Design

2.2.1. Experimental Intervention

The THR and POL groups followed similar exercise formats, but their training intensity distribution (TID) differed. Each group participated in 12 training sessions per week, with details shown in Figure 1.

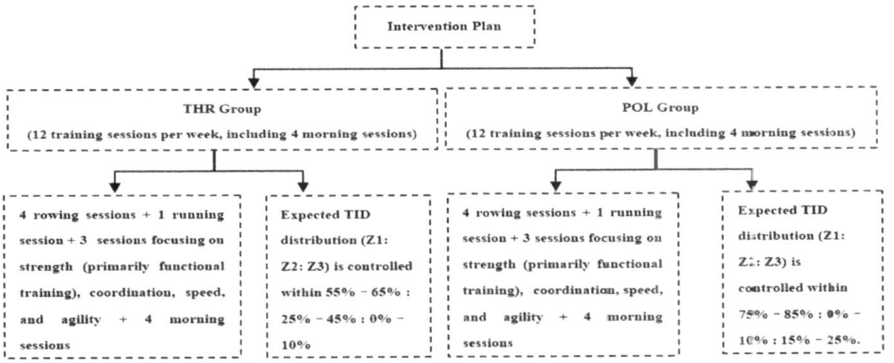

Figure 1. 8-week experimental intervention plan.

2.2.2. Training Intensity Zones and Load Monitoring

Based on the research of Matzka [17], Sylta [18], and others, this study defined the aerobic threshold (Zone 1, Z1) as blood lactate levels below 2.5 mmol/L, marked as the first lactate threshold (LT_1). Blood lactate levels between 2.5 and 4 mmol/L were designated as the transition between aerobic and anaerobic zones (Zone 2, Z2), defined as LT_1 to LT_2. Blood lactate levels of ≥4 mmol/L were categorized as above the anaerobic threshold (Zone 3, Z3), marked as the second lactate threshold (LT_2) [13,19]. Heart rate values corresponding to LT_1 and LT_2 were calculated using the Lactate Analysis Plugin (LAP) [20]. Polar watches (Polar Pacer, Finland) and the Firstbeat Training and Recovery Monitoring System (Firstbeat, Finland) were used to monitor TID continuously. The Time-in-Zone (TIZ) method was employed to calculate TID, focusing solely on heart rate data from endurance activities such as running and rowing on ergometers. Strength training, speed training, coordination exercises, warm-ups, and recovery sessions were identical in both groups and excluded from the TID calculations. To enhance comparability, the experimental design was modeled on studies by Treff et al. [13] and Chiang et al. [21], ensuring relative equivalence in average time spent, exercise distance, energy expenditure, and Training Impulse (TRIMP) between the three intensity zones for both groups [22,23].

2.2.3. Training Load Calculation Standards

Considering that the Firstbeat Training and Recovery Monitoring System may experience data loss when monitoring heart rate during outdoor running sessions, potentially leading to an underestimation of training intensity distribution, this study used Polar

Pacer heart rate data as the primary reference for TID calculations. To provide a more comprehensive and accurate reflection of athletes' physiological responses during each training session, all heart rate data from the entirety of each session, including warm-ups and cool-downs, were included in the statistics. If an athlete missed one or more complete training sessions within a week, their data for that week were excluded from the analysis [20].

2.3. Testing Protocol

Before and after the 8-week training period, participants underwent a 2 km maximal rowing ergometer test. The detailed testing protocol was as follows: Prior to testing, athletes performed a warm-up consisting of 4 laps of slow jogging (approximately 2 min and 30 s per lap), followed by 5 min of joint mobilization and skeletal muscle activation exercises. Next, they completed one 80% effort sprint over 40 m, two 90% effort sprints over 40 m, and one 100% effort sprint over 10 m. After this, a 10 min sport-specific warm-up at a power output of approximately 200 W was conducted to further mobilize the joints. Following the warm-up, athletes rested for 10 min with access to appropriate hydration. Once ready, the ergometer resistance was set to the appropriate level, and the test commenced upon a unified signal from the coach, with participants giving their maximal effort to complete the 2 km rowing test.

2.4. Blood and Urine Indicators

To align with the research objectives, blood samples were collected twice, before and after the 8-week training period. Prior to blood collection, all athletes were instructed to avoid high-intensity physical activity for 48 h. Specifically, at 6:30 a.m., a dedicated nurse from a tertiary hospital collected 5 ml of venous blood from each participant using standard collection tubes. The samples were allowed to clot at room temperature for 1 h, followed by centrifugation at 4 °C at 1500× g for 10 min. Serum was then transferred to clean centrifuge tubes, snap-frozen in liquid nitrogen for 30 s, and stored in a −80 °C freezer for subsequent testing [24].

Additionally, to enhance the accuracy of the test results, the following precautions were taken during sampling: Athletes were required to avoid staying up late and refrain from eating or drinking after 10 p.m. the night before the test. To ensure the athletes were in a fully rested state, all participants were instructed to sit or lie down quietly for at least 20 min before blood collection. For consistency, pre-test blood samples were stored at −80 °C and analyzed alongside post-test samples using the same batch of reagents. This study was approved by the Ethics Committee of the Beijing Sport University Exercise Science Laboratory (Approval Number: 2023009H), and all testing and training procedures adhered strictly to the guidelines of the World Medical Association Declaration of Helsinki.

2.5. Metabolomics Sequencing and Analysis

Polar metabolites were extracted from 50 μL of serum using 200 μL of ice-cold methanol containing 8 μg of phenylhydrazine for α-keto acid derivatization. The samples were vortexed and incubated at −20 °C for 1 h. Subsequently, samples were incubated at 4 °C at 800× g for 15 min. After incubation, samples were centrifuged at 4 °C at 600× g for 10 min, and the clean supernatant was transferred to a new tube and dried using a centrifugal evaporator in H_2O mode. The dried extracts were then reconstituted in 120 μL of 5% acetonitrile aqueous solution, and the supernatant was collected for analysis.

2.5.1. Sample Detection Conditions and Quality Control (QC) Methods

Chromatography conditions: Metabolite separation was performed using an ACQUITY UPLC BEH Amide 1.7 μm, 2.1 × 100 mm column (Waters, Dublin, Ireland) for normal-phase chromatography. An ACQUITY UPLC HSS T3 1.8 μm, 3.0 × 100 mm column (Waters, Dublin, Ireland) was used for reverse-phase chromatography.

2.5.2. Mass Spectrometry Conditions

Analysis was conducted using a 5600 Triple TOF Plus high-resolution tandem mass spectrometer (AB Sciex, Singapore) with an electrospray ionization (ESI) source in negative ion mode at a voltage of −4.5 kV. The evaporator temperature was set to 500 °C, with drying gas (nitrogen) and nebulizer gas (nitrogen) pressures both at 50 psi and a curtain gas (nitrogen) pressure of 35 psi. The scan range was m/z 60–800. The data acquisition mode for metabolites included TOF full-scan for primary mass spectrometry and Information-Dependent Acquisition (IDA) for secondary mass spectrometry, with a collision energy setting of (−)30 ± 15 eV.

2.5.3. Quality Control (QC) Methods

QC samples were prepared by pooling equal volumes (10 μL) of all individual samples, followed by even distribution into aliquots. QC samples were run before and after the main sample queue and at intervals of every 10 samples. The ionization signal of internal standards in QC samples was monitored to ensure stability (within a 20% variation limit) and retention time consistency (within 0.05 min). An average correlation coefficient of 0.99 confirmed high data consistency and quality in LC-MS analyses.

2.6. Data Analysis

All data were analyzed using SPSS 20.0 and are presented as mean ± standard deviation (SD). A two-way repeated measures ANOVA was used to compare the changes in the 2 km maximal rowing performance indicators before and after the intervention for the THR and POL groups. A p-value of < 0.05 was considered statistically significant.

Data acquisition and processing were conducted using AnalystTF 1.7.1 software (AB Sciex, Concord, ON, Canada). MarkerView 1.3 (AB Sciex, Concord, ON, Canada) was employed to extract peak area, mass-to-charge ratio, and retention time from the primary mass spectrometry data, generating a two-dimensional data matrix with isotope peaks filtered. PeakView 2.2 (AB Sciex, Concord, ON, Canada) was utilized to extract MS/MS data and annotate ion information through comparisons with the Metabolites database, HMDBMETLIN, and reference standards, identifying metabolite IDs. The identified IDs were matched to corresponding ions in the primary mass spectrometry matrix. Custom R-based scripts were applied for statistical and pathway analysis. The peak areas of endogenous metabolites were normalized to those of structurally similar isotope-labeled standards for quantification. For endogenous metabolites without isotope-labeled counterparts, the optimal internal standard was chosen using an automated algorithm based on the normalized minimum coefficient of variation [25]. Differential metabolites were screened using paired sample t-tests ($p < 0.05$), and those with fold change (FC) $< 1/1.5$ or FC > 1.5 were considered significant [26]. Differential metabolites were subjected to Kyoto Encyclopedia of Genes and Genomes (KEGG) pathway enrichment and topology analysis, with significant enrichment determined by $p < 0.05$. Perform a two-factor repeated measures ANOVA to compare the changes in endurance performance indicators before and after intervention between the THR group and the POL group.

3. Results

3.1. Training Intensity Distribution

Statistical results of training load during the experimental period (Table 2) indicated that the THR group's training intensity distribution (TID) was 72%:24%:4%, meeting the criteria for the threshold model. The POL group's TID was 78%:8%:14%, with a polarization index (PI) greater than 2, aligning with the standards of the polarized model [13]. No statistically significant differences were observed between the two groups in terms of total distance, total calorie expenditure, and training impulse (TRIMP), indicating relatively equivalent training loads for both groups.

Table 2. Overview of training load statistics for THR and POL groups.

Group	Total Training Duration (h)	Z1 Training Duration (h)	Z2 Training Duration (h)	Z3 Training Duration (h)	Training Intensity Distribution	Total Distance (km)	Total Calories (kcal)	TRIMP	PI
THR	105.89 ± 7.25	76.22 ± 7.52	25.75 ± 2.49	3.91 ± 3.03	72%:24%:4%	2259.00 ± 261.99	53744.67 ± 3876.16	6775.43 ± 1168.77	—
POL	103.72 ± 11.25	80.78 ± 11.05	8.90 ± 1.95	14.05 ± 2.04	78%:8%:14%	2252.13 ± 129.16	50522.75 ± 6774.66	6578.38 ± 520.12	2.11 ± 0.12

Note: All data were obtained from Polar Flow and represent average values over the 8-week training period. TRIMP denotes training impulse, while PI (polarization index) represents the polarization index. "—" indicates that data for this group is not available.

3.2. The 2 km Maximal Rowing Performance

Results from two-way repeated measures ANOVA indicated that after the 8-week training intervention, there was no significant interaction effect between time and group for the 2 km ergometer test ($F = 1.620$, $p = 0.216$). However, the main effect of time was significant ($F = 172.0$, $p < 0.0001$), while the main effect of the group was not significant ($F = 0.2055$ $p = 0.6548$). Post-hoc paired comparisons revealed that both groups showed a gradual and significant improvement in their 2 km maximal rowing performance after the 8-week training ($p < 0.001$), with post-test performance significantly better than pre-test performance ($p < 0.001$), with details shown in Figure 2.

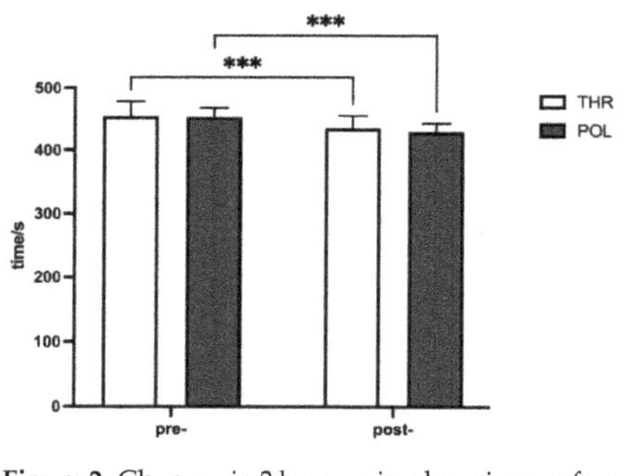

Figure 2. Changes in 2 km maximal rowing performance. Compared to before the 8-week intervention, *** $p < 0.0001$.

3.3. Quality Control Results for Metabolomics

As shown in Figure 3A, the principal component (PC) score plot, including QC samples, reveals a dense clustering of QC samples, indicating stable signals throughout the detection process and high data quality. The data from six QC samples were log10-transformed, and pairwise comparisons were conducted to produce a QC sample correlation matrix. In Figure 3B, the tight linear distribution of all scatter points demonstrates a high degree of consistency between the two QC samples, with corresponding correlation coefficients exceeding 0.99, indicating strong data consistency and high quality. Figure 3C shows that the relative standard deviations (RSD) of the internal standard metabolites were all below 10%, indicating stable instrument signals and high data quality during testing.

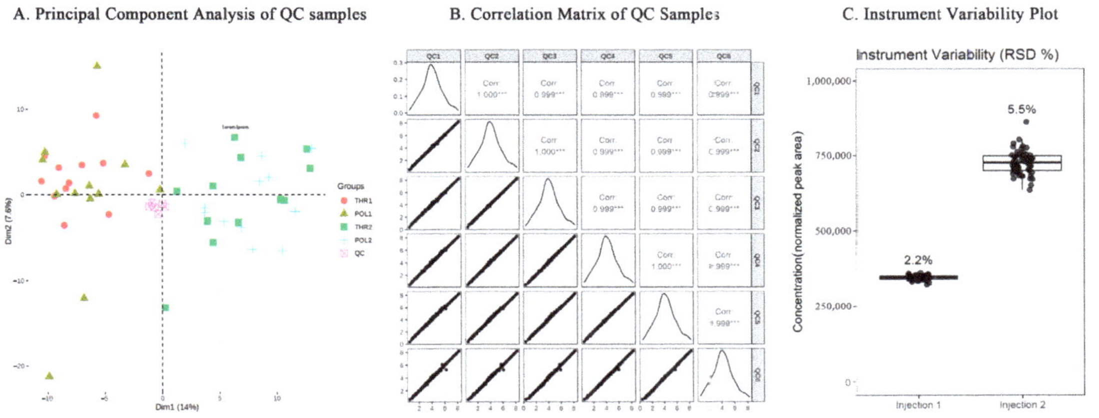

Figure 3. Quality control results of metabolomics experiment data. Note: THR1 denotes the pre-test for the threshold training group, THR2 represents the post-test for the threshold training group, POL1 denotes the pre-test for the polarized training group, and POL2 represents the post-test for the polarized training group; this notation applies throughout (*** represents a high degree of consistency between the data in the two QC samples, reflecting high data quality).

3.4. Principal Component Analysis

As shown in Figure 4A,B, the sample scatter plots for the THR and POL groups during the pre- and post-8-week training intervention at rest exhibit a clear separation trend along the first principal component (PC1) and second principal component (PC2). These results indicate substantial biological metabolic differences in both groups before and after the training intervention.

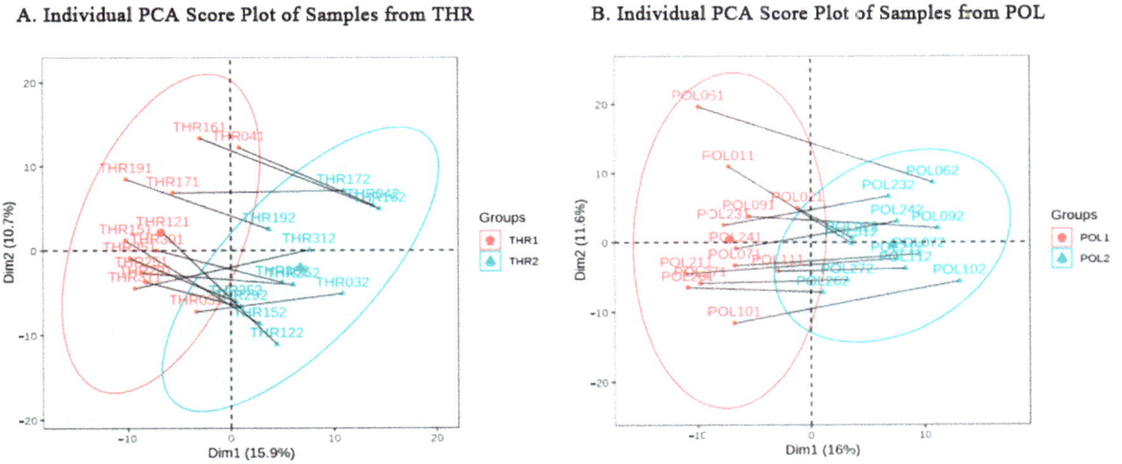

Figure 4. PCA score plot of samples from the THR and POL groups before and after intervention.

3.5. *Differential Metabolite Analysis*

The differential metabolite analysis results for the THR and POL groups are detailed in Figure 5A,B. Metabolites were selected based on criteria of $p < 0.05$ and FC > 1.5, or $p < 0.05$ and FC < 1/1.5 [26]. Compared to pre-intervention levels, the THR group exhibited a total of 35 significantly upregulated metabolites after the 8-week training intervention. The top five upregulated metabolites were 12S-HHT, citric acid, oxidized glutathione, dihydromorphine-6-glucuronide, and daidzein, with respective fold changes of 93.62, 15.48, 6.38, 4.03, and 3.92. The POL group, after 8 weeks of training intervention, showed 38 significantly upregulated metabolites, with the top five being 12S-HHT, citric acid, oxidized glutathione, daidzein, and genistein, with fold changes of 150.23, 45.23, 9.26, 3.95, and 3.19, respectively. The THR group also demonstrated 20 significantly downregulated metabolites after the 8-week intervention. The top five downregulated metabolites were sucrose, arachidonic acid, 12-methyltridecanoic acid, histidylserine, and an isomer of mimosine, with fold decreases of 5.51, 4.50, 2.86, 2.65, and 2.60, respectively. The POL group exhibited 22 significantly downregulated metabolites, with the top 5 being arachidonic acid, 12-methyltridecanoic acid, galactose, L-tryptophan, and L-serine, with respective fold decreases of 5.10, 3.50, 3.10, 2.97, and 2.86.

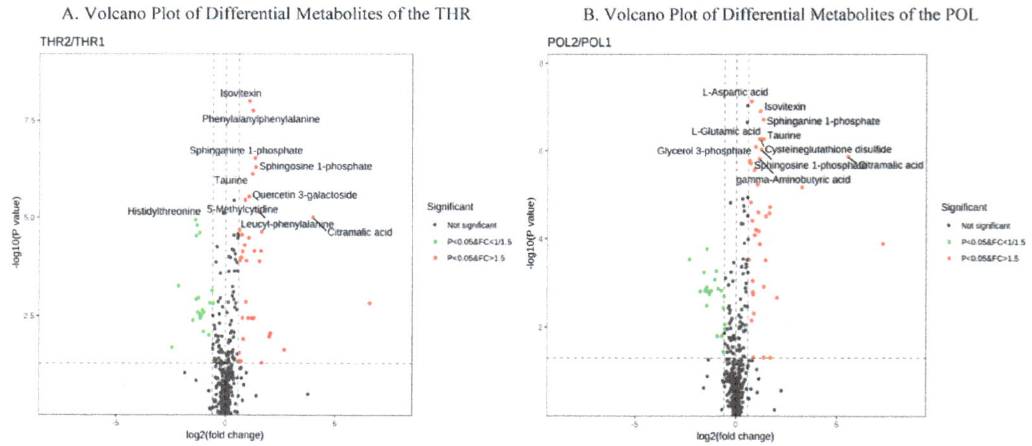

Figure 5. Volcano Plot of Differential metabolites for the THR and POL groups before and after intervention.

3.6. *Differential Metabolic Pathway Enrichment Analysis*

The differential metabolic pathway analysis results for the THR and POL groups are illustrated in Figure 6A,B. Pathway enrichment analysis of the THR group's differential metabolites after the 8-week training intervention revealed four significantly enriched pathways: pyruvate metabolism, glycine, serine, and threonine metabolism, glutathione metabolism, and sphingolipid metabolism. Similarly, analysis of the POL group's differential metabolites identified four significantly enriched pathways: sphingolipid metabolism, glycine, serine, and threonine metabolism, aminoacyl-tRNA biosynthesis, and glutathione metabolism. Detailed results are provided in Table 3.

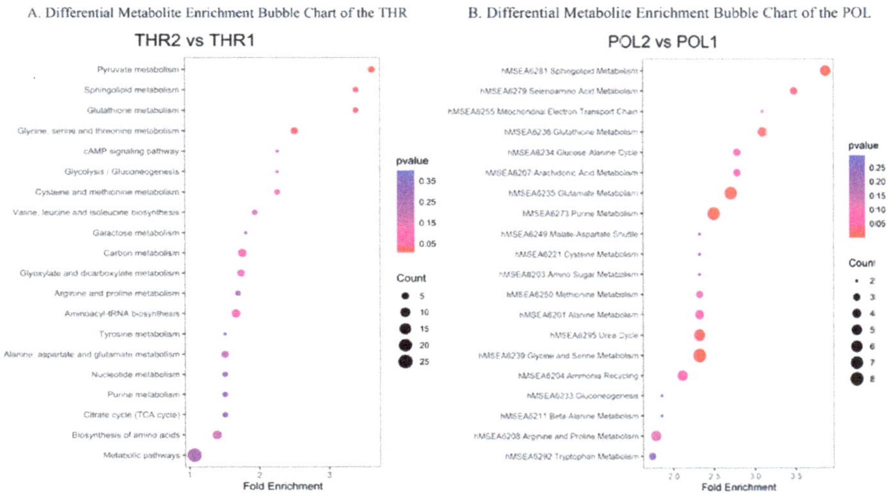

Figure 6. Bubble plot of metabolic pathway enrichment analysis of differential metabolites in the THR and POL groups before and after intervention. Note: Each bubble in the bubble plot represents a metabolic pathway. The size and position of the bubble along the horizontal axis indicate the pathway's impact factor in the topological analysis, with larger values representing greater impact. The color and position of the bubble along the vertical axis represent the enrichment analysis *p*-value (expressed as the negative natural logarithm, $-\log(p)$). Darker bubble colors indicate smaller *p*-values, signifying a higher degree of enrichment.

Table 3. Overview of key metabolic pathways and metabolites before and after 8 weeks of training in the THR Group.

No.	Pathway Name	Metabolites	*p*-Value	Enrichment Fold
1	Pyruvate Metabolism	Fumaric acid, L-Lactic acid, L-Malate, Pyruvic acid	0.0085	3.60
2	Glycine, Serine, and Threonine Metabolism	Glycine, L-Serine, L-Threonine, Pyruvic acid, 2-Ketobutyric acid	0.025	2.50
3	Glutathione Metabolism	Glycine, L-Glutamic acid, Oxidized Glutathione	0.034	3.38
4	Sphingolipid Metabolism	L-Serine, Sphingosine 1-phosphate	0.034	3.38

Note: The table lists the metabolic pathways and metabolites that were significantly altered in the Threshold Training (THR) group following eight weeks of training, along with their respective *p*-values and enrichment fold changes.

3.7. Analysis of Shared Differential Metabolites

As shown in Figure 7A, after the 8-week training intervention, 46 differentially expressed metabolites were found to be shared between the THR and POL groups. Additionally, the THR group exhibited nine uniquely expressed differential metabolites, while the POL group had 14 uniquely expressed differential metabolites. As illustrated in Figure 7B, the key metabolic pathways commonly enriched for both the THR and POL groups before and after the 8-week intervention include sphingolipid metabolism, glutathione metabolism, and glycine, serine, and threonine metabolism.

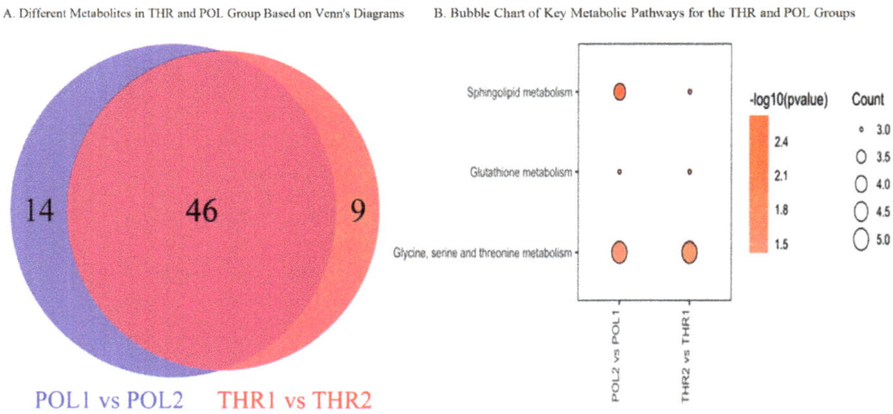

Figure 7. Venn diagram of differential metabolic pathways and bubble plot of key metabolic pathways for the THR and POL groups.

4. Discussion

4.1. Aerobic Endurance Performance

This study found no significant differences in pre-test 2 km maximal rowing performance between the THR and POL groups, indicating comparable baseline levels of sport-specific aerobic endurance. After an 8-week intervention, both the THR and POL groups showed significant improvements in their 2 km ergometer rowing performance, with no significant differences between the groups, suggesting that both training models effectively enhance aerobic endurance performance in adolescent athletes, with comparable effects on aerobic performance. These findings align with the results of Neal [12] and Treff et al. [13]. Neal et al. observed similar outcomes in cycling, where 12 elite male cyclists completed six weeks of either THR or POL training in a randomized crossover design with a four-week washout period between protocols. Both training models effectively increased average power output over a 40 km trial, with similar improvements between groups. Treff et al. [13] examined 14 elite male rowers (aged 20 $\pm$ 2 years) over 11 weeks of THR and POL training and found comparable improvements in 2 km rowing performance.

4.2. Shared Metabolites Enhancing Aerobic Capacity

Results demonstrated significant changes in blood metabolite profiles following 8 weeks of THR and POL training in adolescent male rowers. Endurance, or fatigue resistance, improvements are often reflected in the ability to maintain specific loads or movement quality for longer periods, with increased anti-fatigue capacity or reduced recovery times serving as important indicators of enhanced endurance. Exercise requires coordinated responses from the cardiovascular and respiratory systems to minimize stress on metabolic organs and meet increased energy demands [5]. Improvements in aerobic capacity represent a chronic adaptation involving cardiovascular, respiratory, skeletal muscle, metabolism, and body composition changes. This adaptation encompasses energy metabolism, fatigue and recovery, oxidative stress response, neurometabolism, amino acid metabolism, cell damage and repair, and signaling pathways. Consequently, improvements in energy metabolism, fatigue resistance, recovery, antioxidation, nutritional supplementation, anti-aging, and anti-inflammatory processes enhance aerobic endurance.

4.2.1. Regulating Energy Metabolism and Increasing Energy Reserves

The levels of citric acid in both the THR and POL groups increased by 15.48-fold and 45.23-fold, respectively, following 8 weeks of endurance training, indicating a positive correlation with aerobic endurance. This suggests that both high-intensity and moderate-to-low-intensity training stimulate citric acid metabolism, potentially marking it as an indicator of improved aerobic endurance. The citric acid cycle plays a central role in the oxidation of metabolic substrates and ATP production through oxidative phosphorylation. In a normal physiological state, approximately 75% of cardiac ATP is derived from fatty acids, with the remainder from glucose and lactate. Additionally, the heart can utilize pyruvate, amino acids, and ketone bodies as metabolic substrates, primarily through mitochondrial β-oxidation of fatty acids and glycolysis of glucose in the cytoplasm, followed by mitochondrial pyruvate oxidation. These pathways converge to produce acetyl-CoA, which enters the TCA cycle, generating reduced equivalents (NADH+, FADH+) and GTP, which are converted to ATP through the electron transport chain [27]. As the initial compound in the TCA cycle, citric acid converts carbohydrates, fats, and proteins into carbon dioxide, releasing energy [7]. Increased resting levels of citric acid in athletes indicate enhanced aerobic energy contribution and improved mitochondrial TCA cycle function, suggesting enhanced aerobic metabolic activity [28,29].

L-glutamate, L-malic acid, and other amino acids showed significant upregulation, while L-leucine, L-threonine, and L-alanine were significantly downregulated. These metabolites, classified as glucogenic amino acids, can be converted to glucose via gluconeogenesis for energy production. Glycine, another glucogenic amino acid, can be converted to pyruvate, entering the TCA cycle for aerobic oxidation. These changes indicate enhanced glucose metabolism, a hallmark of improved aerobic capacity. This aligns with previous studies, such as Gao Huan et al. [28], who examined metabolic responses in eight adolescent rowers (age: 16.18 $\pm$ 1.34 years) following 4 weeks of THR training (TID: 74%:26%:0). Glutamate, glycine, and alanine were identified as key markers of aerobic metabolism. Additionally, increased levels of arachidonic acid, a precursor for endocannabinoid synthesis, activate endocannabinoid signaling, modulating metabolism to accumulate energy reserves, reduce glucose utilization in skeletal muscle, and promote hepatic fat synthesis, thereby reflecting enhanced exercise capacity [30].

4.2.2. Reduce Oxidative Stress and Improve Metabolic Environment

Intense or prolonged exercise accelerates substance and energy metabolism, increasing free radical levels. Elevated reactive oxygen species stimulate vagal afferent fibers in the ventricles, reducing aerobic endurance. Glutathione and taurine effectively scavenge intracellular free radicals, reducing oxidative stress and mitigating lipid peroxidation [31]. Taurine supplementation enhances myocardial superoxide dismutase activity, increasing myocardial contractility and promoting post-exercise heart rate recovery [31]. Enhanced free radical scavenging improves the body's internal environment and, consequently, aerobic capacity.

This study showed significant upregulation of 12S-HHT in both the THR and POL groups following 8 weeks of endurance training, with increases of 93.62-fold and 150.23-fold, respectively, potentially serving as a new target for evaluating aerobic endurance improvement. 12S-HHT, an enzymatic product of prostaglandin H2, originates from COX-mediated arachidonic acid metabolism and exhibits protective cellular effects, including nitric oxide regulation, cellular morphology maintenance, and cell flow regulation, enhancing immunity, anti-inflammatory, and antioxidant functions [32].

Daidzein and genistein also exhibit cardioprotective effects. Daidzein reduces arterial sclerosis, mitigates hypoxia and arrhythmia, and modestly lowers blood pressure, benefiting cardiovascular health. Mechanisms by which daidzein improves aerobic capacity may include coronary artery dilation, reduced cardiovascular resistance, and lower myocardial oxygen consumption. Daidzein also reduces thromboxane A2 (TXA2) and prostaglandin-I-2 (PGI2), inhibiting platelet aggregation, reducing blood viscosity, and improving microcirculation. Similarly, genistein, a flavonoid, mitigates arsenic trioxide-induced cardiac toxicity, suppresses endothelial cell inflammation, modulates lipid metabolism, and prevents cardiac fibrosis. Its protective mechanisms may involve inhibiting mitochondrial apoptosis proteins, activating Akt (protein kinase B) and ERK1/2 (extracellular signal-regulated kinases), and enhancing protein expression [33].

4.2.3. Promoting Erythropoiesis and Enhancing Fatigue Resistance

Erythrocytes, the most abundant blood cells, serve as crucial oxygen carriers. Their number is an essential indicator of aerobic capacity. Taurine, oxidized glutathione, and 12S-HHT may play key roles in promoting erythropoiesis. Under fatigue or overtraining, erythrocyte morphology may be altered, while taurine enhances iron absorption in the gut, stabilizes erythrocyte membranes, and improves cell structure [31]. Oxidized glutathione regulates aerobic capacity by converting to reduced glutathione via glutathione reductase, preventing hemoglobin oxidation and reducing erythrocyte oxidative damage. Additionally, Okuno et al. [34] found that 12S-HHT is associated with cytochrome P450 enzymes regulating erythrocyte synthesis, and its upregulation promotes erythropoiesis, contributing to improved aerobic endurance. Quintas et al. [35] revealed significant changes in γ-aminobutyric acid (GABA) levels in elite female soccer players over a season using urine metabolomics. As the primary inhibitory neurotransmitter in the central nervous system, GABA plays a crucial role in balancing neuronal excitation and inhibition, promoting sleep and fatigue resistance.

4.3. *Differential Metabolites Enhancing Aerobic Capacity*

As described above, after the 8-week endurance training intervention, the THR group uniquely expressed nine metabolites, including dihydromorphinone-6-glucuronide, 4-vinylphenol sulfate, hydroxycaprylic acid, L-malic acid, 4-ethylphenyl sulfate, mimetic acid, 7,8-dihydrobiopterin, heptadecanoic acid, and sucrose. Pathway analysis showed that the THR group's unique metabolites were mainly enriched in pyruvate metabolism, which is closely related to the tricarboxylic acid (TCA) cycle. This suggests that the THR training model improves aerobic capacity primarily through enhancing energy metabolism.

Similarly, the POL group uniquely expressed a total of 14 metabolites after the 8-week endurance training intervention, including genistein, O-phosphoethanolamine, 3-indolepropionic acid, unknown sulfated steroids, hydroquinone sulfate, L-aspartic acid, C16 sphingosine-1-phosphate, β-citryl-L-glutamate, butyric acid, allantoin, glutaric acid, 4-(but-2-yl)cyclohexane-1-carboxylic acid, valproic acid, and 2-aminoheptanoic acid. Pathway analysis revealed that the POL group's unique metabolites were enriched in aminoacyl-tRNA biosynthesis, a pathway associated with skeletal muscle protein synthesis. This molecular response perspective further suggests that the POL model's improvement of aerobic capacity is more inclined towards peripheral mechanisms, which may be related to the body's differential activation of energy metabolism systems under varying intensity stimuli [10].

In summary, both the THR group and the POL group have both shared and differential metabolites, with the number of common metabolites significantly outnumbering the differential ones. This suggests that the two endurance training models share common pathways but also have their respective unique aspects in enhancing aerobic endurance.

4.4. Limitations

Although this study identified potential metabolic targets for reflecting aerobic capacity, further validation through larger sample sizes is needed. Future research should focus on targeted metabolomics for key metabolic organs such as the heart and skeletal muscle to enhance detection specificity and provide more scientifically robust evidence. Given the predominance of aerobic endurance data in cardiovascular disease-related databases, future studies should emphasize systematic integration and correlation analysis of similar or functionally related metabolites to advance understanding in this domain.

5. Conclusions

After 8 weeks of training using the THR model (TID of 72%:24%:4%) and the POL model (TID of 78%:8%:14%), the two groups exhibited a high number of commonly expressed metabolites and shared enriched pathways. This indicates that both training models enhance aerobic endurance through similar physiological mechanisms, which may be attributed to a comparable proportion of Z1 intensity, relatively equivalent training loads, and a high overall training volume in both experimental groups. Furthermore, both the THR group and the POL group exhibited numerous commonly expressed metabolites and some differential metabolites, suggesting that the two endurance training models share common pathways but also have distinct aspects in enhancing aerobic endurance.

The mechanisms by which the commonly expressed metabolites enhance aerobic capacity primarily involve regulating energy metabolism, increasing energy reserves, reducing oxidative stress, protecting myocardial cells, promoting erythropoiesis, and improving fatigue resistance.

Our study, for the first time, identified metabolic biomarkers in adolescent rowing athletes after lactate threshold training and polarized training, providing a new theoretical foundation and experimental evidence for revealing the biological mechanisms underlying the enhancement of aerobic capacity. Specifically, the study found that certain metabolites in adolescent rowing athletes, which are associated with nutritional supplements, exhibited significant physiological functions and possessed potential as nutritional supplements after an 8-week endurance training program.

Author Contributions: Writing original draft preparation, F.K. and M.Z. Investigation, J.Z. and S.Y. Software, X.P., C.P. and D.L. Writing review and editing, J.M. and L.Z. All authors have read and agreed to the published version of the manuscript.

Funding: This work was supported financially by grants from the China Fundamental Research Funds for the Central Universities (Beijing Sport University File No. 2024YDXL001 and No. 2023024) and the National Key Research and Development Program under grant number 2022YFC3600201.

Institutional Review Board Statement: This research has been performed in accordance with the Declaration of Helsinki and was approved by the Medical Ethics Committee of Beijing Sport University (protocol code 2023009H). Written informed consent was obtained from all participants included in the study.

Informed Consent Statement: Informed consent was obtained from all subjects involved in the study.

Data Availability Statement: Data are contained within the article.

Acknowledgments: We thank all those who volunteered to participate in the test. This study did not involve animal experiments.

Conflicts of Interest: The authors declare no conflicts of interest.

References

1. Khoramipour, K.; Sandbakk, Ø.; Keshteli, A.H.; Gaeini, A.A.; Wishart, D.S.; Chamari, K. Metabolomics in exercise and sports:a systematic review. *Sports Med.* **2022**, *52*, 547–583. [CrossRef] [PubMed]
2. Benjamin, D.I.; Brett, J.O.; Both, P.; Benjamin, J.S.; Ishak, H.L.; Kang, J.; Kim, S.; Chung, M.; Arjona, M.; Nutter, C.W.; et al. Multiomics reveals glutathione metabolism as a driver of bimodality during stem cell aging. *Cell Metab.* **2023**, *35*, 472–486.e6 [CrossRef] [PubMed]
3. Nicholson, J.K.; Lindon, J.C.; Holmes, E. 'Metabonomics': Understanding the metabolic responses of living systems to patho physiological stimuli via multivariate statistical analysis of biological NMR spectroscopic data. *Xenobiotica* **1999**, *29*, 1181–1189 [CrossRef] [PubMed]
4. Nicholson, J.K.; Wilson, I.D. Opinion:understanding 'global' systems biology:metabonomics and the continuum of metabolism *Nat. Rev. Drug Discov.* **2003**, *2*, 668–676. [CrossRef]
5. Ussher, J.R.; Elmariah, S.; Gerszten, R.E.; Dyck, J.R.B. The emerging role of metabolomics in the diagnosis and prognosis o cardiovascular disease. *J. Am. Coll. Cardiol.* **2016**, *68*, 2850–2870. [CrossRef]
6. Carrard, J.; Guerini, C.; Appenzeller-Herzog, C.; Infanger, D.; Königstein, K.; Streese, L.; Hinrichs, T.; Hanssen, H.; Gallart-Ayala H.; Ivanisevic, J.; et al. The metabolic signature of cardiorespiratory fitness: A systematic review. *Sports Med.* **2022**, *52*, 527–546 [CrossRef]
7. Nemkov, T.; Cendali, F.; Stefanoni, D.; Martinez, J.L.; Hansen, K.C.; San-Millán, I.; D'alessandro, A. Metabolic signatures o performance in elite World Tour professional male cyclists. *Sports Med.* **2023**, *53*, 1651–1665. [CrossRef]
8. Casado, A.; González-Mohíno, F.; González-Ravé, J.M.; Foster, C. Training periodization, methods, intensity distribution, and volume in highly trained and elite distance runners:A Systematic Review. *Int. J. Sports Physiol. Perform.* **2022**, *17*, 820–833 [CrossRef]
9. Pérez, A.; Ramos-Campo, D.J.; Freitas, T.T.; Rubio-Arias, J.; Marín-Cascales, E.; Alcaraz, P.E. Effect of two different intensity distribution training programmes on aerobic and body composition variables in ultra-endurance runners. *Eur. J. Sport Sci.* **2019** *19*, 636–644. [CrossRef]
10. Pan, X.; Zhao, C.; Chen, J.; Sui, H.; Yan, Y. Application Prospect of Multi-omics in Sports Science: Sportomics. *J. Beijing Sport Uni* **2023**, *46*, 52–63.
11. Seiler, K.S.; Kjerland, G.Ø. Quantifying training intensity distribution in elite endurance athletes: Is there evidence fo an"optimal"distribution? *Scand. J. Med. Sci. Sports* **2006**, *16*, 49–56. [CrossRef] [PubMed]
12. Neal, C.M.; Hunter, A.M.; Brennan, L.; O'Sullivan, A.; Hamilton, D.L.; DeVito, G.; Galloway, S.D.R. Six weeks of a polarized training-intensity distribution leads to greater physiological and performance adaptations than a threshold model in trained cyclists. *J. Appl. Physiol.* **2013**, *114*, 461–471. [CrossRef] [PubMed]
13. Treff, G.; Winkert, K.; Sareban, M.; Steinacker, J.M.; Sperlich, B. The polarization-index: A simple calculation to distinguish polarized from non-polarized training intensity distributions. *Front. Physiol.* **2019**, *10*, 707. [CrossRef] [PubMed]
14. Filipas, L.; Bonato, M.; Gallo, G.; Codella, R. Effects of 16 weeks of pyramidal and polarized training intensity distributions in well-trained endurance runners. *Scand. J. Med. Sci. Sport* **2022**, *32*, 498–511. [CrossRef] [PubMed]
15. Burnley, M.; Bearden, S.E.; Jones, A.M. Polarized training is not optimal for endurance athletes. *Med. Sci. Sports Exerc.* **2022**, *54* 1032–1034. [CrossRef]
16. Foster, C.; Casado, A.; Esteve-Lanao, J.; Haugen, T.; Seiler, S. Polarized training is optimal for endurance athletes. *Med. Sci. Sport Exerc.* **2022**, *54*, 1028–1031. [CrossRef]
17. Matzka, M.; Leppich, R.; Sperlich, B.; Zinner, C. Retrospective analysis of training intensity distribution based on race pace versus physiological benchmarks in highly trained sprint kayakers. *Sports Med. Open* **2022**, *8*, 1–12. [CrossRef]
18. Sylta, Ø.; Tønnessen, E.; Seiler, S. Do elite endurance athletes report their training accurately? *Int. J. Sports Physiol. Perform.* **2014** *9*, 85–92. [CrossRef]
19. Lucía, A.; Sánchez, O.; Carvajal, A.; Chicharro, J.L. Analysis of the aerobic-anaerobic transition in elite cyclists during incremental exercise with the use of electromyography. *Br. J. Sports Med.* **1999**, *33*, 178–185. [CrossRef]
20. Farrell, J.W.; Dunn, A.; Cantrell, G.S.; Lantis, D.J.; Larson, D.J.; Larson, R.D. Effects of group running on the training intensity distribution of collegiate cross-country runners. *J. Strength Cond Res.* **2021**, *35*, 2862–2869. [CrossRef]
21. Chiang, T.-L.; Chen, C.; Lin, Y.-C.; Chan, S.-H.; Wu, H.-J. Effect of polarized training on cardiorespiratory fitness of untrained healthy young adults: A randomized control trial with equal training impulse. *J. Sports Sci. Med.* **2023**, *22*, 263–272. [CrossRef] [PubMed]
22. Foster, C.; Florhaug, J.A.; Franklin, J.; Gottschall, L.; Hrovatin, L.A.; Parker, S.; Doleshal, P.; Dodge, C. A new approach to monitoring exercise training. *J. Strength Cond. Res.* **2001**, *15*, 109–115. [PubMed]
23. Plews, D.J.; Laursen, P.B.; Kilding, A.E.; Buchheit, M. Heart-rate variability and training-intensity distribution in elite rowers. *Int. J. Sports Physiol. Perform.* **2014**, *9*, 1026–1032. [CrossRef] [PubMed]

24. Zimmermann, M.; Sauer, U.; Zamboni, N. Quantification and mass isotopomer profiling of α-keto acids in central carbon metabolism. *Anal. Chem.* **2014**, *86*, 3232–3237. [CrossRef] [PubMed]
25. Tian, H.; Ni, Z.; Lam, S.M.; Jiang, W.; Li, F.; Du, J.; Wang, Y.; Shui, G. Precise metabolomics reveals a diversity of aging-associated metabolic features. *Small Methods* **2022**, *6*, e2200130. [CrossRef]
26. Mukherjee, K.; Edgett, B.A.; Burrows, H.W.; Castro, C.; Griffin, J.L.; Schwertani, A.G.; Gurd, B.J.; Funk, C.D. Whole blood transcriptomics and urinary metabolomics to define adaptive biochemical pathways of high-intensity exercise in 50–60 year old masters athletes. *PLoS ONE* **2014**, *9*, e92031. [CrossRef]
27. Campelj, D.; Philp, A. NAD^+ therapeutics and skeletal muscle adaptation to exercise in humans. *Sports Med* **2022**, *52*, 91–99. [CrossRef]
28. Gao, H.; Meng, Z.; Li, T.; Liang, S.; Wang, Y.; Gao, B. The metabolic response to aerobic endurance training in adolescent rowers. *CHN Sport Sci. Techn.* **2021**, *57*, 29–37.
29. Busquets-García, A.; Bolaños, J.P.; Marsicano, G. Metabolic messengers: Endocannabinoids. *Nat. Metab.* **2022**, *4*, 848–855. [CrossRef]
30. Woo, A.Y.H.; Waye, M.M.Y.; Tsui, S.K.W.; Yeung, S.T.W.; Cheng, C.H.K. Andrographolide up-regulates cellular-reduced glutathione level and protects cardiomyocytes against hypoxia/reoxygenation injury. *J. Pharmacol. Exp. Ther.* **2008**, *325*, 226–235. [CrossRef]
31. Gawryluk, J.W.; Wang, J.-F.; Andreazza, A.C.; Shao, L.; Young, L.T. Decreased levels of glutathione, the major brain antioxidant, in post-mortem prefrontal cortex from patients with psychiatric disorders. *Int. J. Neuropsychopharmacol.* **2011**, *14*, 123–130. [CrossRef] [PubMed]
32. Das, U.N. Essential fatty acids and their metabolites in the pathobiology of inflammation and its resolution. *Biomolecules* **2021**, *11*, 1873. [CrossRef] [PubMed]
33. Hu, W.-S.; Lin, Y.-M.; Ho, T.-J.; Chen, R.-J.; Li, Y.-H.; Tsai, F.-J.; Tsai, C.-H.; Day, C.H.; Chen, T.-S.; Huang, C.-Y. Genistein suppresses the isoproterenol-treated H9c2 cardiomyoblast cell apoptosis associated with P-38, Erk1/2, JNK, and NFκB signaling protein activation. *Am. J. Chin. Med.* **2013**, *41*, 1125–1136. [CrossRef] [PubMed]
34. Okuno, T.; Yokomizo, T. Metabolism and biological functions of 12(S)-hydroxyheptadeca-5Z, 8E, 10E- trienoic acid. *Prostaglandins Other Lipid Mediat.* **2021**, *152*, 106502. [CrossRef]
35. Quintas, G.; Reche, X.; Sanjuan-Herráez, J.D.; Martínez, H.; Herrero, M.; Valle, X.; Masa, M.; Rodas, G. Urine metabolomic analysis for monitoring internal load in professional football players. *Metabolomics* **2020**, *16*, 45. [CrossRef]

metabolites

Article

Based on Sportomics: Comparison of Physiological Status of Collegiate Sprinters in Different Pre-Competition Preparation Periods

Pengyu Fu [1,†], Xiaomin Duan [2,†], Yuting Zhang [1], Xiangya Dou [3] and Lijing Gong [4,*]

1 Department of Physical Education, Northwestern Polytechnical University, Xi'an 710072, China; fupy@nwpu.edu.cn (P.F.)
2 Shaanxi Institute of Sports Science, Xi'an 710065, China; hedyduan666@gmail.com
3 College of Life Science, Northwestern Polytechnical University, Xi'an 710072, China
4 Key Laboratory of Exercise and Physical Fitness, Ministry of Education, Beijing Sport University, Beijing 100084, China
* Correspondence: lijing.gong@bsu.edu.cn
† These authors contributed equally to this work and should be considered co-first authors.

Abstract: This study aimed to assess the impact of pre-competition training by comparing the differences of collegiate sprinters in physiological state between strengthening and tapering training period by sportomics and combining their sport performance. Fifteen collegiate sprinters were investigated or tested on their body composition, dietary habits, energy expenditure, sleep efficiency, heart rate and respiratory rate during training, blood (blood cells, biochemical and immune markers) and urine (urinalysis), gut microbiome distribution, microbiome and blood metabolites, and their functions during the strengthening (Group A) and tapering training period (Group B) prior to competing in the national competitions. We found that 26.67% of sprinters achieved personal bests (PB) after the competition. The limb skeletal muscle mass and lymphocyte ratio of male sprinters in Group B were lower than those in Group A, and the serum creatine kinase (CK) level was higher than Group A ($p < 0.05$). The levels of serum CK, interleukin-6 (IL-6), interleukin-1β (IL-1β), and urine-specific gravity (SG) of the two groups were higher than the upper limit of the reference value. The detection rates of urine white blood cell (WBC) and protein in Group B were higher than those in Group A. The gut microbiome health index (GMHI) of Group A was higher than that of Group B, and the microbial dysbiosis index was lower than that of Group B. The ratio of *Firmicutes/Bacteroidota* (F/B) in Group A was higher than that in Group B. There were 65 differential metabolites in the A/B group, and the enriched pathway was mainly the NF-kappa B signaling pathway (up); B/T cell receptor signaling pathway (up); Th1 and Th2 cell differentiation (up); phenylalanine metabolism (up); and growth hormone synthesis, secretion, and action (up). There were significant differences in blood metabolites between the A and B groups, with a total of 89 differential metabolites, and the enriched pathway was mainly the NF-kappa B signaling pathway (up), T cell receptor signaling pathway (up), Th1 and Th2 cell differentiation (up), and glycerophospholipid metabolism (down). In conclusion, the imbalance of the gut microbiome and inflammation and immune-related metabolites of collegiate sprinters during the pre-competition tapering training period may be the cause of their limited sports performance.

Keywords: sprinters; collegiate athletes; pre-competition status; sports performance; gut microbiomes

1. Introduction

The effectiveness of pre-competition training programs is directly impacted by their scientificity and logic. Pre-competition training typically consists of strengthening and tapering the training period. Sprinting is a kind of speed-strength sport, and therefore training volume usually reaches peak three weeks prior to competition and then gradually drops [1]. Collegiate sprinters face pressure from both training and competition, as well as from academics and work, which may significantly differ from professional athletes' psychological and physiological pre-competition conditions [2]. However, there are not many studies on this topic.

Monitoring athletes' pre-competition training plays an important role in developing reasonable training programs. It is known to all that a variety of physiological and biochemical markers, such as body composition, heart rate, respiratory rate, and blood and urine indicators, are frequently used for sports training monitoring. The changes of these markers help coaches and athletes thoroughly understand the physical condition and training effect of athletes by reflecting the nutrition and metabolic health, hydration balance, muscle status, endurance performance, fatigue, and inflammation levels. Although these biomarkers have been widely used, the analysis of their results still faces many challenges: (1) the sensitivity of single biomarkers to detect overtraining is limited, (2) the reference range for different training levels has not been clearly defined, and (3) interindividual variance in absolute values and relative changes in biomarkers [3].

Therefore, systematically assessing the impacts of pre-competition training modifications based on single metabolites and fewer pathways is frequently challenging in traditional investigations. Metabolomics and other omics technologies are applied to investigate the metabolic effects of physical exercise on individuals, which is known as "sportomics", having gained popularity in the field of exercise physiology [4,5]. We attempted to apply "sportomics" to assess, as is rarely documented, the effects of various pre-competition preparation phases on collegiate sprinters' condition.

Collegiate athletes train at a lesser volume and intensity than professional athletes. Conventional physiological and biochemical markers might not be accurate predictors of pre-competition training conditions. Drawing on "sportomics", metabolomics and microbiomics were used in our study. Changes in metabolic levels can be detected by metabolomics in real time, within hours, or even days [6], which have been used to identify metabolites at various competition and preparation nodes in competitive sports [7,8]. "Providing fuel for microbes" could develop into a key tactic for enhancing sports performance. This concept has made the microbiome a research hotspot in competitive sports [9]. The microbiome has been employed in some recent research to monitor athlete training, and studies show that this is a highly significant tool for understanding possible inflammatory hazards and creating individualized training regimens [10].

We hypothesized that gut microorganisms and their metabolites, as well as blood metabolites during the pre-competition strengthening and tapering training period, are different, which, combined with the variations in physiological and biochemical markers, can help predict sports performance. This study aims to evaluate the rationality of the pre-competition training program of collegiate sprinters by searching for more sensitive pre-competition training monitoring markers and provide theoretical guidance for the formulation of scientific and effective pre-competition preparation programs.

2. Materials and Methods

2.1. Participants and Groups

This study involved 15 collegiate sprinters, who were training at Northwestern Polytechnical University, at the period of preparation for the 21st National Collegiate Track and Field Championships of China in 2023. The sprinters were 21.20 $\pm$ 1.74 years old and had been accepted into the professional training for 6.8 $\pm$ 2.18 years. All were national second-level or above (sports technology level of China) athletes, and 10 of them were male and 5 were female. The sprinters shared a lifestyle (lived and ate together) and pre-competition training before the competition. Subjects were excluded if they (1) were unable to train because of a recent injury; (2) used antibiotics and probiotic supplements in the previous 6 months; and (3) had gastrointestinal diseases, such as diarrhea and constipation.

We monitored the physiological conditions of the sprinters during the strengthening and the tapering training period and divided these two periods into Group A (n = 15) and Group B (n = 15) (Figure 1).

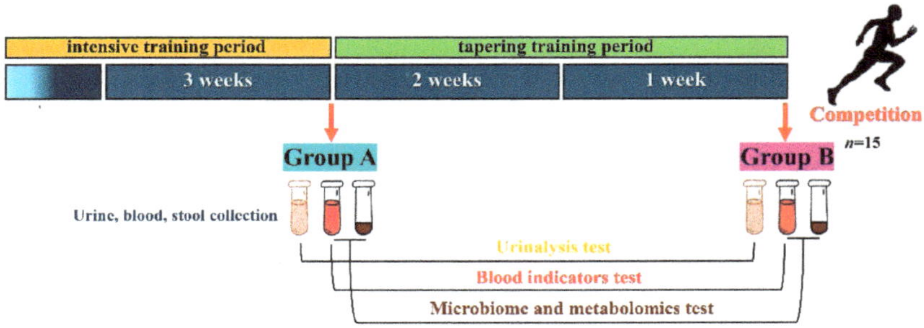

Figure 1. Conditions for collecting biological samples from sprinters.

The training plans for two periods are shown in Table 1. All sprinters completed informed permission forms after fully understanding the research purpose and program with the coaches' assistance and cooperation.

Table 1. Two-period training plan.

Times	Mon.	Tues.	Wed.	Thur.	Fri	Sat.
			Strengthening Training Period (Group A)			
Warm-up exercise	Warm-up and run for 10 min, cool-down for 10 min					
Formal training	100 m run × 1 set + 60 m run × 2 sets + 30 m run × 2 sets	300 m run × 5 sets	Upper and lower limb and back strength training	200 m run × 4 sets + 150 m run × 4 sets	Hurdle jumping, hurdle running, hip and ankle mobility training	Full speed special athletics training
Cooling-down exercise	Stretching for 15 min					
			Tapering training period (Group B)			
Warm-up exercise	Warm-up and run for 10 min, cool-down for 10 min					
Formal training	100 m run × 1 set + 60 m run × 1 set + 30 m run × 1 set	300 m run × 2 sets	Upper and lower limb and back strength training	-	-	-
Cooling-down exercise	Stretching for 15 min					

2.2. Body Composition

The direct digital detectors (GE PRODIGY) were used to test the body composition of sprinters in a fasting state.

2.3. Dietary Survey

The 24 h recall method (http://www.ihatepsm.com/blog/24-hour-recall-questionnaire-method, access date: 1 July 2023) was used to record the daily dietary intake (including food types and estimated weights of three meals and snacks) by the BooHee Health mobile application (APP) (this APP can estimate the weight of food and calculate energy expenditure) one week before the competition [11].

2.4. Energy Expenditure and Sleep Efficiency

The three-axis accelerometers (ActiGraph wGT3X-BT, ActiGraph, Pensacola, FL, USA) were worn on the left wrist throughout the day to record daily physical activity energy expenditure, average activity intensity, and sleep efficiency.

2.5. The Test of Heart Rate and Respiratory Rate

The Zephyr physiological status monitoring training system was worn on the chest to record the heart rate and respiratory rate during strength training and special athletics training.

2.6. Blood and Urine Indicators

A total of 10 mL of fasting venous blood was drawn to test blood cells and biochemical and immunological indicators; 15 mL of the first mid-morning urine was collected, and routine urine tests were performed using a urine analyzer. The above tests were completed in the University Hospital.

2.7. Gut Microbiome Sequencing and Analysis

The middle section of the first stool in the morning was collected in a 2 mL collection tube and frozen with liquid nitrogen. The Illumina PE300/PE250 platform (Illumina, San Diego, CA, USA) was used for the high-throughput sequencing of the 16S rRNA V3-V4 region in stool samples. Bioinformatic analysis of the gut microbiome was carried out using the Majorbio Cloud platform (https://cloud.majorbio.com, access date: 1 September 2023). Bacterial community characterization index and composition were analyzed.

2.8. Metabolomics Sequencing and Analysis

A total of 50 mg of stool and 100 μL of urine samples were weighed and extracted with 400 μL of extraction solution, respectively. The liquid chromatography–mass spectrometry system (LC-MS/MS) analysis of samples was conducted on a Thermo UHPLC-Q Exactive HF-X system equipped with an ACQUITY HSS T3 column at Majorbio Bio-Pharm Technology Co., Ltd. (Shanghai, China). The pretreatment of LC/MS raw data was performed by Progenesis QI (Waters Corporation, Milford, CT, USA) software, and a three-dimensional data matrix in CSV format was exported. The data matrix obtained by searching the database was uploaded to the Majorbio cloud platform (https://cloud.majorbio.com, access date: 1 September 2023) for data analysis. The R package "ropls" (Version 1.6.2) was used to perform principal component analysis (PCA) and orthogonal least partial squares discriminant analysis (OPLS-DA). The metabolites with variable importance in the projection (VIP) > 1, $p < 0.05$, were determined as significantly different metabolites based on the VIP obtained by the OPLS-DA model and the p-value generated by Student's t test. Differential metabolites among two groups were mapped into their biochemical pathways through metabolic enrichment and pathway analysis based on the Kyoto Encyclopedia of Genes and Genomes (KEGG) database (http://www.genome.jp/kegg/, access date: 1 October 2023).

2.9. Statistical Analysis

All the data were analyzed using SPSS20.0 and expressed as mean ± standard deviation (SD). The paired t-test was used to determine the statistical difference of body composition, energy metabolism, sleep efficiency, heart and respiratory rate, and blood and urine markers between the two groups. The gut microbiome characterization index between the two groups was analyzed using the Wilcoxon rank sum test (false discovery rate (FDR) correction). $p < 0.05$ was considered statistically significant (Cohen's d reflects the effect size).

PCA was converted to unit variance (UV) with a confidence level of 0.95. Student's t-test was used to compare the two groups in the differential metabolite analysis. The screening conditions were p-value < 0.05, VIP-pred-OPLS-DA > 1, and the fold change (FC) > 1. Seven-fold cross validation was used to ascertain the VIP using OPLS-DA/PLS-DA as the supervised model. KEGG pathway enrichment analysis was performed by relative-betweenness centrality and BH multiple test correction.

3. Results

3.1. Sports Performance and Body Composition

After the competition, four sprinters achieved their personal bests (PB), including three male and one female sprinter. There were no significant differences in body mass index (BMI), body fat percentage, and muscle mass between the two groups in male and female sprinters. The body fat percentage and trunk skeletal muscle mass of male sprinters in Group A increased compared with those in Group B, while the body fat percentage of female sprinters in Group A decreased, and the trunk skeletal muscle mass increased compared with those in Group B. The limb skeletal muscle mass of male sprinters in Group B was significantly lower than Group A ($p < 0.01$, Cohen's d = 0.54), and which in female sprinters in Group B also decreased (Table 2).

Table 2. Body composition of the sprinters (mean ± SD, n = 15).

Indicators (Unit)	Male (n = 10)		Female (n = 5)	
	Group A	Group B	Group A	Group B
Height (cm)	178.75 ± 3.19		169.10 ± 5.32	
Body weight (kg)	68.52 ± 5.25	68.24 ± 5.59	56.88 ± 3.49	56.25±3.32
Body mass index	21.37 ± 1.18	21.31 ± 1.25	19.92 ± 0.57	19.68 ± 0.54
Body fat percentage (%)	10.29 ± 3.64	11.51 ± 2.08	18.60 ± 2.17	17.22 ± 4.32
Skeletal muscle mass (kg)	32.81 ± 2.61	31.50 ± 2.22	24.18 ± 1.58	24.32 ± 1.11
Trunk skeletal muscle mass (kg)	6.68 ± 0.79	6.96 ± 1.05	6.40 ± 0.48	6.90 ± 0.78
Limb skeletal muscle mass (kg)	26.10 ± 2.10	25.04 ± 1.80 **	17.80 ± 1.11	17.42 ± 0.67

Compared with Group A, ** $p < 0.01$.

3.2. Energy Metabolism and Sleep

There were no differences in the three major nutrient energy supplies and energy intakes nor in sleep efficiency between the two groups. The energy expenditure of male sprinters in Group B was significantly lower than Group A ($p < 0.001$, Cohen's d = 2.18), but there were no differences in females (Figure 2).

3.3. Heart Rate and Respiratory Rate during Training

The maximum heart rate ($p < 0.05$, Cohen's d = 0.80) and mode of respiratory rate ($p < 0.01$, Cohen's d = 1.04) in Group B were lower than those of Group A in strength training. The mode of heart rate in Group B was higher than that of Group A ($p < 0.05$, Cohen's d = 0.89), while the mode of respiratory rate was lower than that of Group A ($p < 0.01$, Cohen's d = 1.23) in special athletics training (Table 3).

Table 3. Heart rate and respiratory rate during training of the sprinters (mean ± SD, n = 15).

Indicators (Times/min)			Group A	Group B
Strength training	Heart rate	Maximum	182 ± 29	164 ± 14 *
		Average	117 ± 17	113 ± 13
		Mode	115 ± 17	110 ± 13
	Respiratory rate	Maximum	423 ± 11	37 ± 6
		Average	23 ± 3	21 ± 2
		Mode	24 ± 5	20 ± 3 **

Table 3. *Cont.*

Indicators (Times/min)			Group A	Group B
Specific athletics training	Heart rate	Maximum	185 ± 15	188 ± 20
		Average	120 ± 9	128 ± 9
		Mode	110 ± 19	124 ± 12 *
	Respiratory rate	Maximum	43 ± 5	46 ± 5
		Average	23 ± 7	24 ± 3
		Mode	24 ± 5	20 ± 3 **

Compared with Group A, * $p < 0.05$, ** $p < 0.05$.

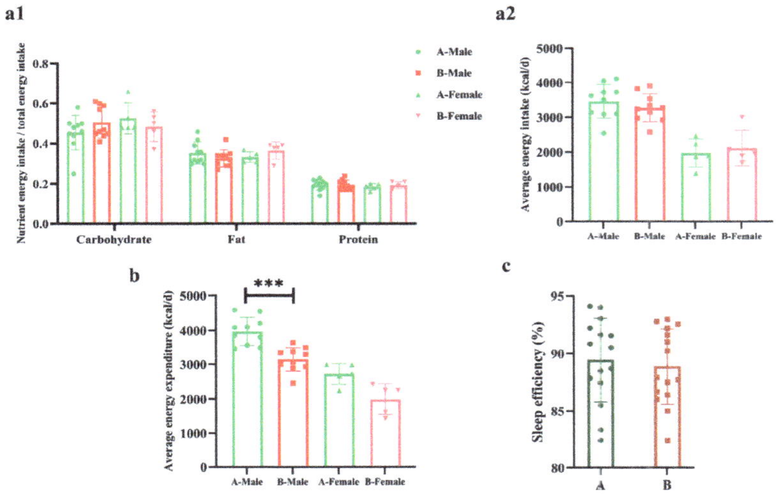

Figure 2. The energy metabolism and sleep efficiency of the sprinters (n = 15). (**a1**) The three major nutrient energy intake ratio; (**a2**) the average daily energy intake; (**b**) the average daily energy expenditure; (**c**) the sleep efficiency. Compared with Group A-Male, *** $p < 0.001$.

3.4. Blood Markers

There was no significant difference in the levels of white blood cell (WBC), immunoglobulin A (IgA), red blood cell (RBC), ferritin, hemoglobin (HGB), mean corpuscular hemoglobin concentration (MCHC), testosterone, cortisol, testosterone/cortisol, blood urea nitrogen (BUN), lactate dehydrogenase (LDH), interleukin-6 (IL-6), or interleukin-1β (IL-1β) between the two groups. The lymphocyte ratio in Group B was lower than in Group A ($p < 0.05$, Cohen's d = 0.99); the creatine kinase (CK) levels (M: $p < 0.05$, Cohen's d = 1.44; F: $p < 0.01$, Cohen's d = 2.71) in Group B were higher than in Group A. The levels of CK, IL-6, and IL-1β were higher than the reference value (Table 4).

Table 4. Blood indicators of sprinters (mean ± SD, n = 15).

Blood Indicators (Unit) (Reference Values)	Implication	Group A		Group B	
		Male (n = 10)	Female (n = 5)	Male (n = 10)	Female (n = 5)
WBC (10^9 / L) (3.69~9.16)	Immune status	5.24 ± 0.89		5.35 ± 1.04	
Lymphocyte ratio (%) (24~48.4)		46.43 ± 6.85		39.20 ± 7.63 *	
IgA (g/L) (0.72~4.29)		2.24 ± 0.70		2.26 ± 0.73	

Table 4. *Cont.*

Blood Indicators (Unit) (Reference Values)	Implication	Group A		Group B	
		Male (n = 10)	**Female (n = 5)**	**Male (n = 10)**	**Female (n = 5)**
RBC (10^{12}/L) (M: 4.3~5.8; F: 3.8~5.1)	Aerobic capacity	5.11 ± 0.39	4.52 ± 0.26	5.08 ± 0.39	4.64 ± 0.25
Ferritin (ng/mL) (M: 21.81~274.66; F: 4.63~204.00)	Aerobic capacity Nutritional status	101.02 ± 29.04	25.33 ± 12.68	123.30 ± 22.43	33.27 ± 16.59
HGB (g/L) (M: 130~175; F: 115~150)	Aerobic capacity Training loads	153.30 ± 8.29	129.20 ± 10.11	152.50 ± 8.72	132.60 ± 9.96
MCHC (g/L) (310~370)		350.10 ± 8.79	336.40 ± 12.10	355.00 ± 9.10	339.40 ± 10.78
Testosterone (ng/mL) (M: 2.49~8.36; F: 0.029~0.481)	Training loads Recovery ability	7.39 ± 1.32	0.48 ± 0.23	6.91 ± 1.56	0.52 ± 0.15
Cortisol (μg/dL) (4.26~24.85)		19.27 ± 4.09	17.66 ± 4.49	17.43 ± 3.35	17.47 ± 2.79
T/C		0.40 ± 0.09	0.03 ± 0.02	0.40 ± 0.10	0.03 ± 0.01
CK (U/L) (M: 38~174; F: 26~140)	Training intensity Muscle injury and recovery	191.76 ± 68.26	217.19 ± 53.62	286.65 ± 63.59 *	370.32 ± 59.07 **
BUN (mmol/L) (2.9~8.2)	Training volume	5.45 ± 1.08		5.11 ± 0.94	
LDH (U/L) (109~245)	Anaerobic capacity	163.23 ± 19.29		167.45 ± 25.49	
IL-6 (pg/mL) (≤7)	Inflammatory response	9.29 ± 5.29		12.87 ± 9.28	
IL-1β (pg/mL) (≤5)		17.28 ± 9.24		20.62 ± 13.20	

Compared with Group A, * $p < 0.05$, ** $p < 0.01$.

3.5. Urine Markers

No glucose (GLU), blood (BLD), nitrite (NIT), urobilinogen (UBG), bilirubin (BIL), or ketone (KET) were detected in the urine of the two groups. The detection rates of urine WBC (reflecting inflammatory status) and protein (PRO) (reflecting training intensity) in Group B were higher than those in Group A. The pH values (reflecting dietary structure) of both groups were slightly acidic. The specific gravity (SG) (reflecting hydration status) levels were higher than the upper limit of the reference value (Table 5).

Table 5. Urine indicators of sprinters (mean ± SD, n = 15).

Urine Indicators		Group A	Group B
WBC	The detection rates	13.33%	35.71%
GLU		(-)	(-)
BLD		(-)	(-)
PRO		6.67%	35.71%
NIT		(-)	(-)
UBG		(-)	(-)
BIL		(-)	(-)
KET		(-)	(-)
pH (3~7.5)	The test value (Reference values)	5.63 ± 0.74	5.62 ± 0.74
SG (1.010~1.025)		1.028 ± 0.003	1.027 ± 0.003

(-) indicates not detected.

3.6. Gut Microbiome Characterization Index and Composition Analysis

The gut microbiome health index (GMHI) of Group A was higher than that of Group B ($p < 0.001$, Cohen's d = 2.04) (Figure 3(a1)). The microbial dysbiosis index (MDI) in Group B was higher than that in Group A ($p < 0.001$, Cohen's d = 2.26) (Figure 3(a2)). At the phylum level, the community composition of the two groups is shown in Figure 3(b1,b2), where the *Firmicutes*/*Bacteroidota* (F/B) ratio in Group A was higher than in Group B.

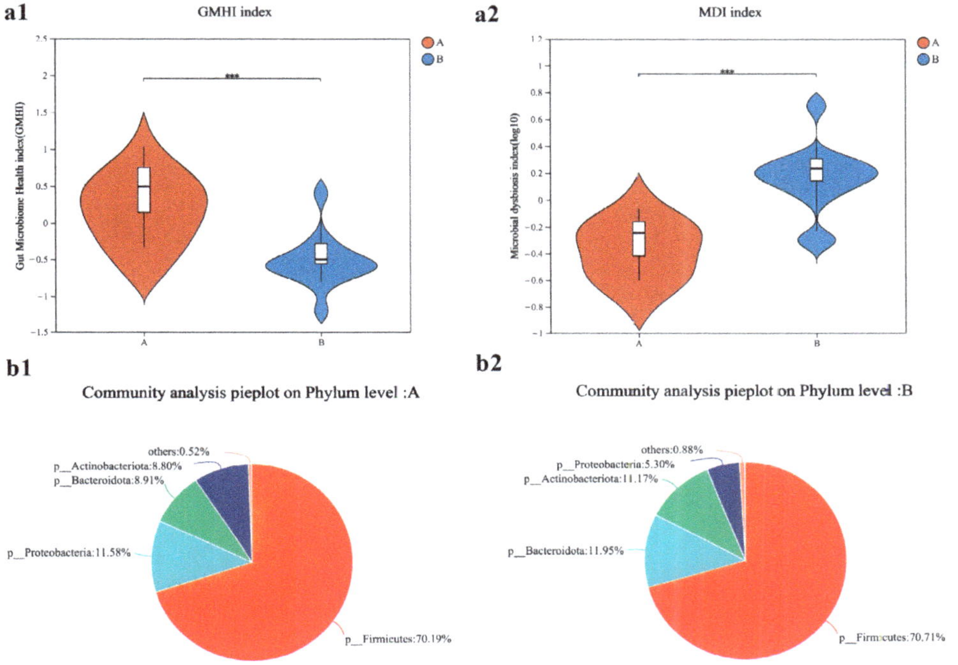

Figure 3. Bacterial community characterization index and composition analysis of sprinters (n = 15). (**a1**) The gut microbiome health index; (**a2**) the microbial dysbiosis index. Compared with Group A, *** $p < 0.001$. The community distribution of Group A (**b1**) and Group B (**b2**). "p" represents phylu.

3.7. Metabolomic Analysis of the Gut Microbiome

PCA analysis results showed that there was little difference in the metabolites of gut microbiome between the two groups (Figure 4a). There were 65 differential metabolites in Group A/B, of which 45 were upregulated and 20 were downregulated (Figure 4b). The top 30 differential metabolites with the highest VIP values are shown in Figure 4c, including anserine (up), etc. (Figure 4c). The pathways enriched by metabolites revealed that the inflammation and immunity related were NF-kappa B signaling pathway (up), B/T cell receptor signaling pathway (up), and Th1 and Th2 cell differentiation(up); amino acid metabolism was phenylalanine metabolism (up); endocrine system was growth hormone synthesis, secretion, and action (up) (Figure 4d).

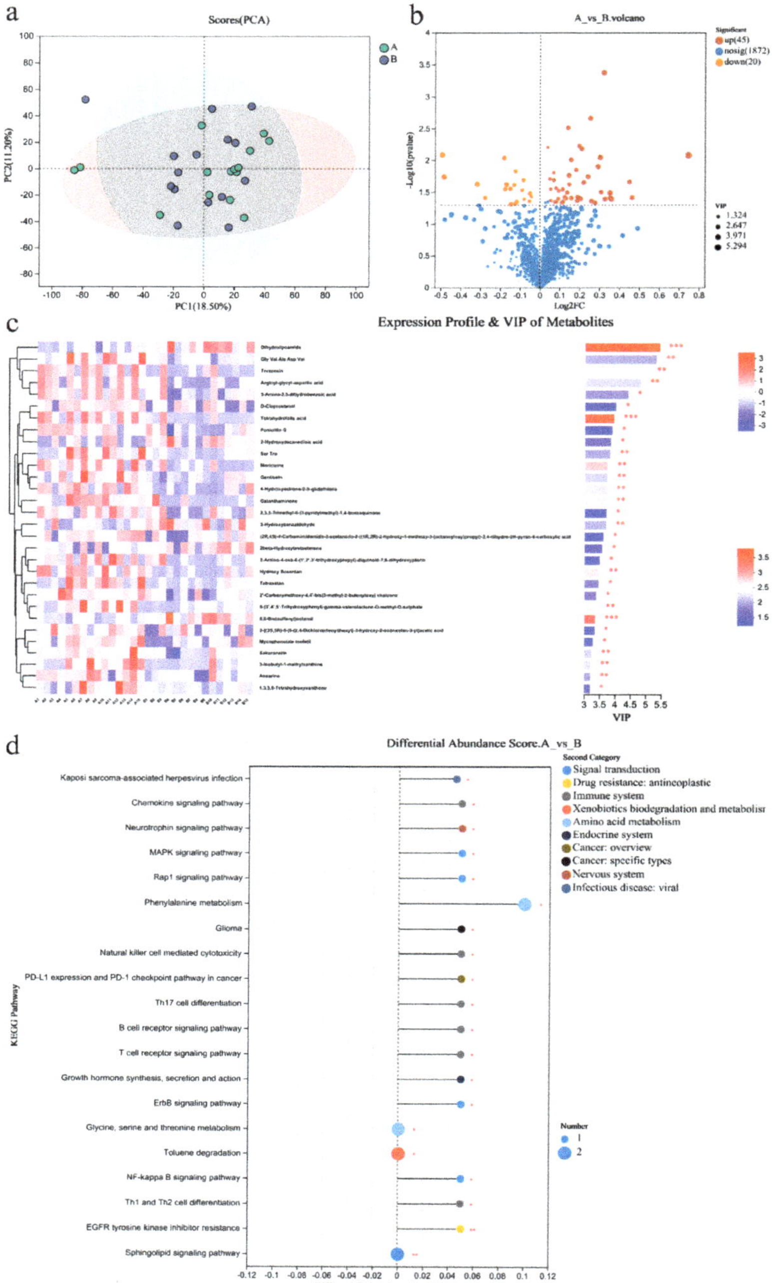

Figure 4. Differential metabolites of the gut microbiome of sprinters and their enriched pathways (n = 15). (**a**) The principal component analysis (PCA); (**b**) the volcano plot of differential metabolites; (**c**) the variable importance in the projection (VIP) analysis chart; (**d**) the differential abundance score plot of the KEGG pathway. Compared with Group A, * $p < 0.05$, ** $p < 0.01$, *** $p < 0.001$.

3.8. Metabolomic Analysis of Blood

PCA analysis results showed that there was a significant difference in the metabolites of blood between the two groups (Figure 5a). There were 89 differential metabolites in Group A/B, of which 40 were upregulated and 49 were downregulated (Figure 5b). The top 30 differential metabolites with the highest VIP values are shown in Figure 5c, including guanosine (down), guanine (down), etc. The pathways enriched by metabolites revealed that the related inflammation and immunity were NF-kappa B (NF-κB) signaling pathway (up), T cell receptor signaling pathway (up), Th1 and Th2 cell differentiation (up); the lipid metabolism was the glycerophospholipid metabolism (down) (Figure 5d).

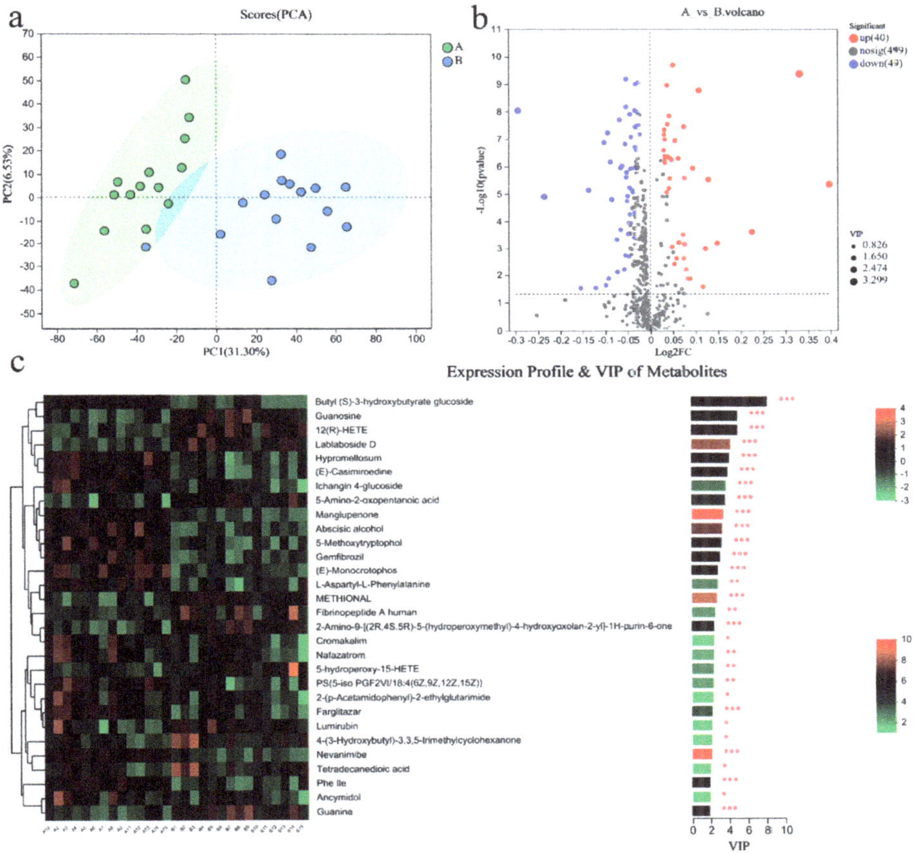

Figure 5. *Cont.*

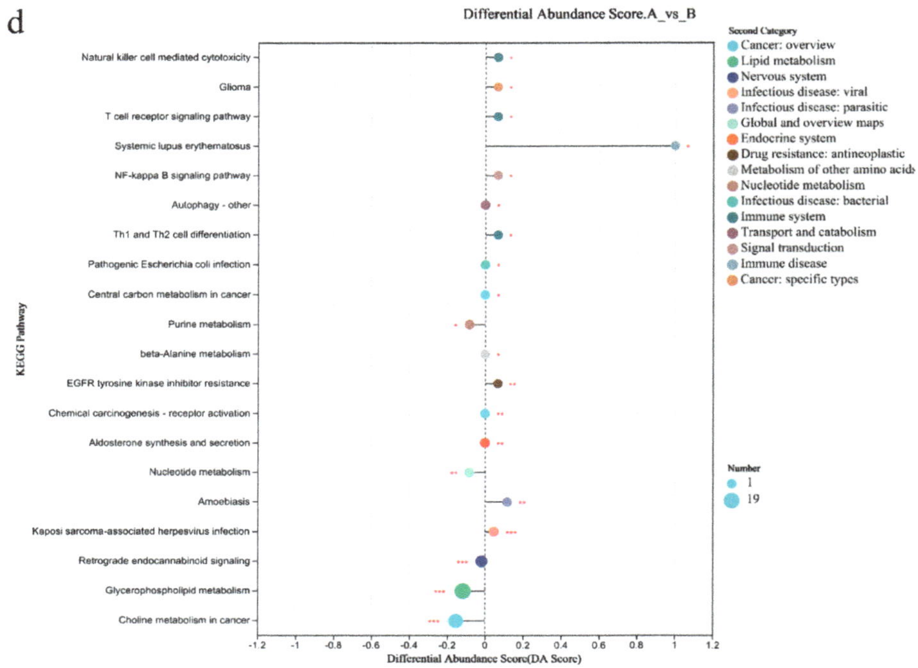

Figure 5. Differential metabolites of the blood of sprinters and their enriched pathways (n = 15). (**a**) The PCA. (**b**) The volcano plot of differential metabolites. (**c**) The VIP analysis chart. (**d**) The differential abundance score plot of the KEGG pathway. Compared with Group A, * $p < 0.05$, ** $p < 0.01$, *** $p < 0.001$.

4. Discussions

This study investigated the energy metabolism, training load, body composition, and blood and urine biochemical markers combined with the gut microbiome and metabolites of collegiate sprinters during the pre-competition strengthening and tapering training period to comprehensively analyze their pre-competition status and predict sports performance.

4.1. Diet, Training Load Adaptation, and Sports Performance

A healthy diet can help improve sports performance [12]. Sprints are characterized by their short duration, rapid pace, and high intensity. It is a typical sport that relies on anaerobic metabolism for energy supply and has high requirements on the speed and strength of athletes. In order to ensure adequate glycogen reserves and muscle mass, sprinters should increase their intake of carbohydrates and proteins before competitions. Pre-competition "overfilling" muscle glycogen is a crucial nutritional strategy for sprinters to enhance their sports performance [13]. A study of endurance runners showed that those who followed a high-carbohydrate diet had better sports performance than those following a high-protein diet [14,15]. Since energy expenditure and macronutrient composition are markers of dietary recommendations for altering body composition [16], we included these two variables to reflect the sprinters' diet in this study. The recommended daily energy intake for males is 4080 kcal and for female is 3190 kcal, and the recommended carbohydrate daily intake is 663 g for males and 518 g for females, based on the weight of our sprinters [17]. We discovered that the sprinters' overall energy intake and carbohydrate intake did not significantly increase between the two pre-competition periods, and the levels were lower than the recommended values: the mean daily energy intakes of male sprinters in Groups A and B were 3466 and 3275 kcal, respectively, while those of female sprinters were 1984 and 2120 kcal, respectively; the daily carbohydrate intakes of male sprinters were 394 and 413 g, respectively, while those of female sprinters were 261 and 251 g, respectively. This could have been one of the reasons for poor sport performance. We speculate that this may be related to the decreased appetite caused by sports fatigue, or it may be related to the fact that collegiate athletes have a relative lack of knowledge about sports nutrition and have not deliberately increased their energy and carbohydrate intake.

Furthermore, sprinters' urine was sub acidic during both periods, and their SG was higher than the test value's upper limit, suggesting that neither group consumed enough water [18].

In our study, the training volume was decreased throughout the tapering training period while the training load remained unchanged from the strengthening training period. Sprinters were less adaptive to training in special athletics training, as evidenced by a decrease in heart rate, which may be related to sports fatigue. This was also reflected in their urine biochemical markers, such as the higher detection rate of urine WBC and protein during the tapering training period. In contrast, sprinters were more adaptive during strength training.

4.2. Gut Microbiome and Sports Performance

The hypothalamic–pituitary–adrenal (HPA) axis can be activated by athletes undergoing high-intensity (over 60% of maximum oxygen consumption, VO_{2max}) and long-duration (above 90 min) training prior to competition, as well as by the tension and anxiety that comes with it [19]. The gut–brain axis, which describes the reciprocal modulation between the gut and brain, is significantly impacted by the activity of the HPA axis. Certain bacterial populations may change as a result of this two-way communication, and the released metabolites may then alter host behavior [20].

We found that sprinters in the tapering training period had an unbalanced gut microbiota, a lower F/B ratio, and somewhat lower sleep efficiency. Because it is positively connected with the levels of short-chain fatty acids (SCFAs) [21] and VO_{2max} [22], the F/B ratio is thought to be a sign of the dynamic balance of the microbiome and is higher in elite athletes [23].

Sports fatigue could be attributed to microbiome imbalance of our sprinters. Combined with the results of blood indicators in this study, it can be found that there was a decline in the lymphocyte ratio of sprinters during the tapering training period. This is an important indication of immunosuppression (EIS). Studies have found that long-term training lasting over 90 min at a level of intensity higher than 65~75% VO_{2max} without enough recovery, might quickly cause exercise-induced EIS, which raises the risk of bacterial and viral infections of athletes [19,24]. Muscle microtrauma and gut microbiome changes caused by high-intensity exercise are both important mechanisms leading to EIS. The serum CK levels considerably increased during the tapering training period in our study, and it was found to be higher than the reference value during both periods. Lack of oxygen and muscle contraction during high-intensity exercise cause damage to the muscle cell membrane; increased permeability; and CK infiltration from the cells into the blood, leading to an increase in serum CK [25]. This leads us to the topic of the relationship between the gut microbiota and skeletal muscle. Through its impact on the gut–muscle axis, the gut microbiota eventually influences muscle function and facilitates the process of muscle synthesis from protein intake [26]. Thus, according to our study, abnormal dietary nutrient intake and microbiome imbalance may be linked to male sprinters' decreased limb muscle mass during the tapering training period.

4.3. Metabolites and Sports Performance

Nowadays, sports science has made extensive use of metabolomics, a potent technique for precisely assessing metabolic pathways at the system level, to investigate metabolic responses that change depending on the type, intensity, and duration of exercise. While the majority of research has concerned individuals with metabolic disorders or those participating in amateur sports, some recent studies have concentrated on the use of metabolomics in professional athletes' training monitoring [27]. It is commonly known that serum or plasma samples make excellent candidate samples for the discovery of biomarkers in metabolomics. In addition, a multitude of microbial metabolic processes found in feces lead to the synthesis or alteration of a broad spectrum of molecules, including vitamins, amino acids, and saturated fat fatty acids (SCFAs) as well as other bioactive molecules like polyphenols and secondary bile acids, which are vital for the regulation of physiological processes and are used to provide energy and necessary nutrients [28].

The pathways enriched by microorganisms and blood metabolites were primarily immunological and inflammatory pathways in our study. It was more evidence that the sprinters were fatigued, especially when the blood levels of IL-6 and IL-1β were higher than the upper limit of the permissible amount [29]. A study has shown that the NF-κB signaling pathway in skeletal muscle can be activated by the body's hypoxic condition during intense exercise, putting the body in an oxidative stress state [30]. Similarly, our study revealed that the level of anserine in Group B of microbial metabolites was downregulated. It has been discovered that anserine can reverse the oxidative damage caused by exercise. Anserine supplements can increase the level of glutathione disulfide (GSSG) and the activity of superoxide dismutase (SOD), preserving the body's antioxidant capability [31]. Furthermore, both microbial and blood metabolites were enriched in T cell immune regulation in this study. One of the most significant biological parameters linked to the risk of viral infection in highly physically active people appears to be the balance between T1 and T2 immune cells. For athletes to experience immunological modulation following intense exercise, T cells are essential. These include Th1 cells, which are gradually activated throughout the recovery phase after intense exercise, and T2 cells, whose IL-6 levels continue to rise and cause Th17 cells to differentiate, function as a pro-inflammatory factor, and eventually contribute to tissue inflammation [32].

In our study, sprinters' pre-competition training also affected the levels of other metabolites, such as guanosine and guanine, which were upregulated in Group B. A study has shown that maximal physical exertion is accompanied by increased degradation of purine nucleotides in muscle, with the products of purine catabolism accumulating in plasma [33]. Glycerophospholipid metabolism was also downregulated in our study. A study found that the blood metabolite levels of multiple pathways related to glycerophospholipids were altered after the competition in collegiate soccer players [34]. A study has also shown that phospholipid-related metabolites such as phosphatidylcholine and phosphatidylserine play an important role in alleviating exercise fatigue [35].

5. Conclusions

Some collegiate sprinters had an unsuitable diet throughout the pre-competition tapering training period and were unable to adjust to the training load, which manifested as gut microbiota imbalance and increased expression of inflammatory and immune-related metabolites. We conclude that changes in microbial and metabolite levels may be a more sensitive way to monitor collegiate athletes training, because their training volume and load are lower than those of professional athletes and may not be sufficient to cause significant changes in the levels of traditional physiological and biochemical indicators. Further investigations into specific metabolic and microbiological indicators relevant to particular sports may be undertaken in the future.

Author Contributions: Writing—original draft preparation, P.F. and X.D. (Xiaomin Duan). Investigation, Y.Z. Software, X.D. (Xiangya Dou). Writing—review and editing, L.G. All authors have read and agreed to the published version of the manuscript.

Funding: This research was funded by Shaanxi Province's "14th Five-Year Plan" Education Planning Project, grant number SGH23Q0269, and Shaanxi Provincial Sports Bureau Scientific Research Project, grant number 2022020.

Institutional Review Board Statement: This research has been performed in accordance with the Declaration of Helsinki and was approved by the Medical Ethics Committee of Northwestern Polytechnical University (protocol code 202302040). Written informed consent was obtained from all participants included in the study.

Informed Consent Statement: Informed consent was obtained from all subjects involved in the study.

Data Availability Statement: Data are contained within the article.

Acknowledgments: We thank all those who volunteered to participate in the test. This study did not involve animal experiments.

Conflicts of Interest: The authors declare no competing interest.

References

1. Malsagova, K.A.; Kopylov, A.T.; Stepanov, A.A.; Kulikova, L.I.; Izotov, A.A.; Yurku, K.A.; Balakin, E.I.; Pustovoyt, V.I.; Kaysheva, A.L. Metabolomic and Proteomic Profiling of Athletes Performing Physical Activity under Hypoxic Conditions. *Sports* **2024**, *12*, 72. [CrossRef] [PubMed]
2. Hamlin, M.J.; Wilkes, D.; Elliot, C.A.; Lizamore, C.A.; Kathiravel, Y. Monitoring Training Loads and Perceived Stress in Young Elite University Athletes. *Front. Physiol.* **2019**, *10*, 34. [CrossRef] [PubMed]
3. Lee, E.C.; Fragala, M.S.; Kavouras, S.A.; Queen, R.M.; Pryor, J.L.; Casa, D.J. Biomarkers in Sports and Exercise: Tracking Health, Performance, and Recovery in Athletes. *J. Strength Cond. Res.* **2017**, *31*, 2920–2937. [CrossRef] [PubMed]
4. Khoramipour, K.; Sandbakk, O.; Hassanzadeh Keshteli, A.; Gaeini, A.; Wishart, D.; Chamari, K. Metabolomics in Exercise and Sports: A Systematic Review. *Sports Med.* **2022**, *52*, 547–583. [CrossRef] [PubMed]
5. Bassini, A.; Cameron, L.C. Sportomics: Building a new concept in metabolic studies and exercise science. *Biochem. Biophys. Res. Commun.* **2014**, *445*, 708–716. [CrossRef]
6. Bongiovanni, T.; Lacome, M.; Fanos, V.; Martera, G.; Cione, E.; Cannataro, R. Metabolomics in Team-Sport Athletes: Current Knowledge, Challenges, and Future Perspectives. *Proteomes* **2022**, *10*, 27. [CrossRef]
7. Pintus, R.; Bongiovanni, T.; Corbu, S.; Francavilla, V.C.; DessÌ, A.; Noto, A.; Corsello, G.; Finco, G.; Fanos, V.; Marincola, F.C. Sportomics in professional soccer players: Metabolomics results during preseason. *J. Sports Med Phys. Fit.* **2020**, *61*, 324–330. [CrossRef]
8. Hudson, J.F.; Phelan, M.M.; Owens, D.J.; Morton, J.P.; Close, G.L.; Stewart, C.E.J.M. "Fuel for the damage induced": Untargeted metabolomics in elite rugby union match play. *Metabolites* **2021**, *11*, 544. [CrossRef]
9. Hughes, R.L.; Holscher, H.D. Fueling Gut Microbes: A Review of the Interaction between Diet, Exercise, and the Gut Microbiota in Athletes. *Adv. Nutr.* **2021**, *12*, 2190–2215. [CrossRef]
10. Li, Y.; Cheng, M.; Zha, Y.; Yang, K.; Tong, Y.; Wang, S.; Lu, Q.; Ning, K. Gut microbiota and inflammation patterns for specialized athletes: A multi-cohort study across different types of sports. *mSystems* **2023**, *8*, e0025923. [CrossRef]
11. Ludbrook, C.; Clark, D. Energy expenditure and nutrient intake in long-distance runners. *Nutr. Res.* **1992**, *12*, 689–699. [CrossRef]
12. Donati Zeppa, S.; Agostini, D.; Gervasi, M.; Annibalini, G.; Amatori, S.; Ferrini, F.; Sisti, D.; Piccoli, G.; Barbieri, E.; Sestili, P.; et al. Mutual Interactions among Exercise, Sport Supplements and Microbiota. *Nutrients* **2019**, *12*, 17. [CrossRef] [PubMed]
13. Pérez-Guisado, J. Athletic performance: Muscle glycogen and protein intake. *Apunt. Sports Med.* **2008**, *43*, 142–151.
14. Furber, M.J.W.; Young, G.R.; Holt, G.S.; Pyle, S.; Davison, G.; Roberts, M.G.; Roberts, J.D.; Howatson, G.; Smith, D.L. Gut Microbial Stability is Associated with Greater Endurance Performance in Athletes Undertaking Dietary Periodization. *mSystems* **2022**, *7*, e0012922. [CrossRef] [PubMed]
15. Naderi, A.; Gobbi, N.; Ali, A.; Berjisian, E.; Hamidvand, A.; Forbes, S.C.; Koozehchian, M.S.; Karayigit, R.; Saunders, B. Carbohydrates and Endurance Exercise: A Narrative Review of a Food First Approach. *Nutrients* **2023**, *15*, 1367. [CrossRef]
16. Aragon, A.A.; Schoenfeld, B.J.; Wildman, R.; Kleiner, S.; VanDusseldorp, T.; Taylor, L.; Earnest, C.P.; Arciero, P.J.; Wilborn, C.; Kalman, D.S.; et al. International society of sports nutrition position stand: Diets and body composition. *J. Int. Soc. Sports Nutr.* **2017**, *14*, 16. [CrossRef]
17. Thomas, D.T.; Erdman, K.A.; Burke, L.M. Nutrition and athletic performance. *Med. Sci. Sports Exerc.* **2016**, *48*, 543–568.
18. Osterberg, K.L.; Horswill, C.A.; Baker, L.B. Pregame urine specific gravity and fluid intake by National Basketball Association players during competition. *J. Athl. Train.* **2009**, *44*, 53–57. [CrossRef]
19. Wegierska, A.E.; Charitos, I.A.; Topi, S.; Potenza, M.A.; Montagnani, M.; Santacroce, L. The Connection Between Physical Exercise and Gut Microbiota: Implications for Competitive Sports Athletes. *Sports Med.* **2022**, *52*, 2355–2369. [CrossRef]
20. Matijašić, M.; Meštrović, T.; Paljetak, H.; Perić, M.; Barešić, A.; Verbanac, D. Gut Microbiota beyond Bacteria-Mycobiome, Virome, Archaeome, and Eukaryotic Parasites in IBD. *Int. J. Mol. Sci.* **2020**, *21*, 2668. [CrossRef]
21. Yamamura, R.; Nakamura, K.; Kitada, N.; Aizawa, T.; Shimizu, Y.; Nakamura, K.; Ayabe, T.; Kimura, T.; Tamakoshi, A. Associations of gut microbiota, dietary intake, and serum short-chain fatty acids with fecal short-chain fatty acids. *Biosci. Microbiota Food Health* **2020**, *39*, 11–17. [CrossRef] [PubMed]

22. Durk, R.P.; Castillo, E.; Márquez-Magaña, L.; Grosicki, G.J.; Bolter, N.D.; Lee, C.M.; Bagley, J.R. Gut Microbiota Compositio Is Related to Cardiorespiratory Fitness in Healthy Young Adults. *Int. J. Sport Nutr. Exerc. Metab.* **2019**, *29*, 249–253. [CrossRef] [PubMed]
23. Han, M.; Yang, K.; Yang, P.; Zhong, C.; Chen, C.; Wang, S.; Lu, Q.; Ning, K. Stratification of athletes' gut microbiota: Th multifaceted hubs associated with dietary factors, physical characteristics and performance. *Gut Microbes* **2020**, *12*, 184299 [CrossRef] [PubMed]
24. Kakanis, M.W.; Peake, J.; Brenu, E.W.; Simmonds, M.; Gray, B.; Hooper, S.L.; Marshall-Gradisnik, S.M. The open window c susceptibility to infection after acute exercise in healthy young male elite athletes. *Exerc. Immunol. Rev.* **2010**, *16*, 119–13 [CrossRef] [PubMed]
25. Baird, M.F.; Graham, S.M.; Baker, J.S.; Bickerstaff, G.F. Creatine-kinase- and exercise-related muscle damage implications fo muscle performance and recovery. *J. Nutr. Metab.* **2012**, *2012*, 960363. [CrossRef]
26. Ticinesi, A.; Lauretani, F.; Tana, C.; Nouvenne, A.; Ridolo, E.; Meschi, T. Exercise and immune system as modulators of intestina microbiome: Implications for the gut-muscle axis hypothesis. *Exerc. Immunol. Rev.* **2019**, *25*, 84–95.
27. San-Millán, I.; Stefanoni, D.; Martinez, J.L.; Hansen, K.C.; D'Alessandro, A.; Nemkov, T. Metabolomics of Endurance Capacity i World Tour Professional Cyclists. *Front. Physiol.* **2020**, *11*, 578. [CrossRef]
28. Rowland, I.; Gibson, G.; Heinken, A.; Scott, K.; Swann, J.; Thiele, I.; Tuohy, K. Gut microbiota functions: Metabolism of nutrient and other food components. *Eur. J. Nutr.* **2018**, *57*, 1–24. [CrossRef]
29. Dupuy, O.; Wafa, D.; Theurot, D.; Bosquet, L.; Dugue, B. An Evidence-Based Approach for Choosing Post-exercise Recover Techniques to Reduce Markers of Muscle Damage, Soreness, Fatigue, and Inflammation: A Systematic Review with Meta-Analysi *Front. Physiol.* **2018**, *9*, 403. [CrossRef]
30. Gallego-Selles, A.; Galvan-Alvarez, V.; Martinez-Canton, M.; Garcia-Gonzalez, E.; Morales-Alamo, D.; Santana, A.; Gonzale Henriquez, J.J.; Dorado, C.; Calbet, J.A.L.; Martin-Rincon, M. Fast regulation of the NF-κB signalling pathway in human skelet muscle revealed by high-intensity exercise and ischaemia at exhaustion: Role of oxygenation and metabolite accumulation. *Redo Biol.* **2022**, *55*, 102398. [CrossRef]
31. Alkhatib, A.; Feng, W.H.; Huang, Y.J.; Kuo, C.H.; Hou, C.W. Anserine Reverses Exercise-Induced Oxidative Stress and Preserve Cellular Homeostasis in Healthy Men. *Nutrients* **2020**, *12*, 1146. [CrossRef] [PubMed]
32. Kostrzewa-Nowak, D.; Nowak, R. Differential Th Cell-Related Immune Responses in Young Physically Active Men after a Endurance Effort. *J. Clin. Med.* **2020**, *9*, 1795. [CrossRef] [PubMed]
33. Dudzinska, W.; Lubkowska, A.; Dolegowska, B.; Safranow, K.; Jakubowska, K. Adenine, guanine and pyridine nucleotides i blood during physical exercise and restitution in healthy subjects. *Eur. J. Appl. Physiol.* **2010**, *110*, 1155–1162. [CrossRef] [PubMed]
34. Vike, N.L.; Bari, S.; Stetsiv, K.; Talavage, T.M.; Nauman, E.A.; Papa, L.; Slobounov, S.; Breiter, H.C.; Cornelis, M.C. Metabolomi response to collegiate football participation: Pre- and Post-season analysis. *Sci. Rep.* **2022**, *12*, 3091. [CrossRef]
35. Wang, C.-C.; Shi, H.-H.; Zhang, L.-Y.; Ding, L.; Xue, C.-H.; Yanagita, T.; Zhang, T.-T.; Wang, Y.-M. The rapid effects of eicosape taenoic acid (EPA) enriched phospholipids on alleviating exercise fatigue in mice. *RSC Adv.* **2019**, *9*, 33863–33871. [CrossRef]

metabolites

MDPI

Article

Serum Metabolomics Reveals Metabolic Changes in Freestyle Wrestlers During Different Training Stages

Xiaonan Li [1,†], Xiangyu Liu [2,†], Jianxing Liu [1], Yinhai Liu [1], Yumei Han [1,*] and Wei Zhang [1,*]

1 School of Physical Education, Shanxi University, Taiyuan 030000, China; lixiaonan@sxu.edu.cn (X.L.)
2 School of Physical Education, Huainan Normal University, Huainan 232000, China; liuxiangyu@hnnu.edu.cn (X.L.)
* Correspondence: hanyumei@sxu.edu.cn (Y.H.); zhangwei6@sxu.edu.cn (W.Z.); Tel.: +86-554-6863665 (Y.H.)
† These authors have contributed equally to this work.

Abstract

Objectives: This study aimed to analyze metabolites changes in elite freestyle wrestlers during three specific training phases—pre-training, peak training, and recovery adjustment—through serum metabolomics analyses and biochemical indicator testing, providing preliminary insights for selecting effective functional assessment metrics. **Methods**: Five male wrestlers (20.40 ± 2.07 years) and five female wrestlers (19.60 ± 0.55 years) were enrolled. Morning fasting venous blood samples were collected before training, at peak training intensity, and after training adjustment and recovery. Serum metabolomic analyses using ^{1}H nuclear magnetic resonance (^{1}H NMR) spectroscopy and assessment of biochemical indicators were performed. **Results**: The metabolomic analysis identified six significantly altered serum biomarkers in male wrestlers and three in females across different training phases. These differential metabolites are primarily implicated in the regulation of energy and amino acid metabolism pathways. Additionally, significant alterations in conventional biochemical indices were observed. **Conclusions**: Metabolomic markers provide a more accurate and comprehensive reflection of metabolic characteristics in freestyle wrestlers, offering a promising complementary approach to traditional biochemical assessments for monitoring physiological states.

Keywords: freestyle wrestlers; serum metabolomics; metabolites

1. Introduction

Freestyle wrestling, as a high-intensity and intermittent combat sport, is characterized by 2 min rounds of intense competition in which athletes must execute various explosive movements within extremely short bursts while maintaining physical combat [1]. Athletes are required to perform various explosive movements dominated by anaerobic energy supply in a single match, while simultaneously maintaining sustained aerobic loads (≥30% VO_2 max for ≥5 min) [2,3]. The energetic demands of such activity involve complex transitions between concentric and eccentric muscle actions, reminiscent of the asymmetric energy expenditure observed between positive and negative mechanical work in gradient locomotion [4]. Furthermore, periodic dehydration and dietary restriction programs that are implemented under weight-class systems may increase the risk of muscle catabolism, electrolyte imbalance, and metabolic homeostasis disruption [5], manifesting as metabolic stress characterized by oxidative stress, lactic acidosis, and glycogen depletion. Traditional training monitoring methods can only reflect localized changes in single metabolic pathways, and it is difficult to reveal the overall metabolic response mechanisms involving multiple systems. Therefore, the introduction of systematic analysis methods is urgently needed. Biochemical markers such as blood lactate, creatine kinase, cortisol, and testosterone are commonly used to monitor athletes' training status. However, these indicators provide isolated snapshots rather than a comprehensive picture of physiological state, making it difficult to reveal the overall metabolic response [6]. Therefore, there is an urgent need to introduce systems-level analytical methods to comprehensively characterize athletes' metabolic adaptation [7].

Metabolomics is an emerging systems biology research method that offers the technical advantages of high sensitivity and high throughput [8]. With rigorously standardized protocols, it enables the qualitative and quantitative analyses of low-molecular-weight metabolites (<1000 Da) in organisms with good reproducibility, allowing a comprehensive representation of changes in metabolic phenotypes during exercise [9]. In recent years, significant progress has been made in the application of metabolomics technology in the field of athletic training [10]. Belhaj et al. noted that this technology can map the molecular landscape of exercise physiology, enabling precise predictions for training load, fatigue recovery, and personalized nutritional interventions [11]. Heaney et al. found that non-targeted metabolomics analyses can reveal dynamic changes in hundreds of metabolites in serum, urine, or saliva, thereby illuminating real-time states of training intensity, fatigue recovery, and energy metabolism [12]. Existing research has primarily focused on metabolic responses in athletes engaged in endurance sports such as running and swimming [13,14], but there remains a significant gap in metabolomics studies concerning high-intensity intermittent combat sports like wrestling, coupled with a lack of dynamic monitoring data across complete training cycles.

Therefore, this study took elite freestyle wrestlers from Shanxi Province as participants, employing ^{1}H NMR metabolomics technology to conduct longitudinal serum analyses over a one-week training period. This comprehensive approach aimed to characterize key metabolic shifts and potential pathway perturbations induced by freestyle wrestling training loads. This not only addresses a gap in systems biology research for combat sports but also provides molecular-level theoretical support for optimizing training cycle design and developing targeted nutritional intervention strategies. It holds practical significance for advancing scientific training in competitive sports.

2. Materials and Methods

2.1. Participants and Ethical Statement

The study included five female freestyle wrestlers and five male freestyle wrestlers, all of whom held athletic titles of First-Level Athlete or higher, as defined by the Athlete Technical Rank Standards issued by the General Administration of Sport of China. The personal basic information of the participants is shown in Table 1. All participants were in good health, had no history of chronic diseases, and were not undergoing any weight loss or weight gain programs. During the experimental period, all participants resided at the training center and were subject to a standardized schedule and dietary management, refrained from taking any medications or sports supplements affecting metabolism, and received no special interventions. The last training session prior to blood sampling was conducted at least 12 h earlier to minimize acute exercise effects. All participants were informed of the testing procedures and objectives and signed informed consent forms. This experimental protocol was reviewed and approved by the Academic Ethics Committee of Shanxi University, and the approved number was SXULL2024110.

Table 1. Basic information of participants.

Sex	Age (Years)	Height (cm)	Weight (kg)
Men	20.40 ± 2.07	180.20 ± 3.83	88.60 ± 15.66
Women	19.60 ± 0.55	164.20 ± 4.02	64.40 ± 4.51

2.2. Sample Collection and Storage

Participants underwent fasting blood sampling at 7:00 AM via the antecubital vein to collect 7 mL of blood at three distinct time points: pre-training (Day 0), peak training (Day 6), and post-training (Day 8) (Figure 1). One tube (2 mL) was placed in an EDTA anticoagulant tube for whole blood analyses. Another tube (5 mL) was placed in a vacuum blood collection tube. After centrifugation at 3000 r/min for 30 min, the serum was collected and aliquoted into two EP tubes. One tube was used for serum biochemical parameter testing, and the other was stored at −80 °C for subsequent metabolomics analyses.

Figure 1. Flow chart of the experimental design.

2.3. Measurement of Biochemical Indicators

ABX automated hematology analyzer (Horiba ABX, Montpellier, France) was used to measure white blood cell count (WBC), red blood cell count (RBC), hemoglobin (HB), hematocrit (HCT), and mean corpuscular volume (MCV). Kits (Biosino Bio-Technology and Science Incorporation, Beijing, China) were employed to detect creatine kinase (CK), blood urea nitrogen (BUN), and testosterone (T). Testing was performed according to the methods specified in the instruction manual.

2.4. Metabolomics Analyses Methods

2.4.1. H-NMR Sample Preparation

After thawing serum samples at 4 °C, 500 μL of serum was aliquoted into a microcentrifuge tube. Then, 350 μL of D_2O was added, followed by centrifugation at 4 °C and 18,000× *g* for 20 min. A total of 600 μL of the supernatant was transferred to an NMR tube for spectral acquisition using a Bruker 600 MHz spectrometer (Bruker BioSpin GmbH, Rheinstetten, Germany) at 25 °C.

2.4.2. Spectral Data Processing and Analyses

The serum NMR spectra were processed using MestReNova 6.1.0 software. After Fourier transformation, chemical shifts were calibrated to the trimethylsilylpropanoic acid (TSP) peak (δ 0.00). A standardized iterative protocol was applied for manual baseline and phase correction to ensure flat baselines and symmetrical peaks, minimizing bias. The spectra were segmented into bins of 0.04 ppm across the range of 0.5–10.0 ppm. To avoid interference from the residual water signal, the region between 4.5 and 4.7 ppm was excluded. To validate multivariate models, R^2Y, Q^2, and permutation tests (200 permutations) were performed, confirming model robustness and minimal overfitting. Finally, the areas of all integral values are normalized and exported to Excel, yielding a matrix of each segment and its corresponding integral area value. Metabolite identification followed the Metabolomics Standards Initiative (MSI) guidelines.

2.5. Statistical Analyses

Data are expressed as mean ± SD. Statistical analyses were performed using SPSS 18.0 (LLC, San Diego, CA, USA). Graphs were generated with GraphPad Prism 7.0 (IBM, Chicago, IL, USA). Multivariate statistical analyses were conducted with SIMCA-P 14.1 (Umetrics, Malmö, Sweden). Principal component analysis (PCA) was first applied to observe intrinsic clustering and outliers. Subsequently, partial least squares-discriminant analyses (PLS-DA) and orthogonal partial least squares-discriminant analyses (OPLS-DA) were employed to enhance group separation and identify differentially expressed metabolites. Metabolites with variable importance in projection values (VIP) greater than 1 and with p-values less than 0.05 were considered statistically significant.

3. Results

3.1. Results of Biochemical Indicator Tests

Table 2 presents the biochemical indicator test results for athletes of different sexes across various training phases. Male wrestlers showed significantly higher concentrations of CK and BU during peak training periods compared to pre-training, with the concentrations of CK significantly decreasing post-training. Compared with pre-training, female wrestlers showed significantly increased serum concentrations of CK ($p < 0.05$) and significantly decreased concentrations of T ($p < 0.05$) during the peak training period. After adequate rest, their CK concentrations significantly decreased following post-training.

Table 2. Blood biochemical parameters of athletes by sex.

Parameter	Male Wrestlers			Female Wrestlers		
	Pre-Training	Peak Training	Post-Training	Pre-Training	Peak Training	Post-Training
WBC (×109 g/L)	5.40 ± 0.88	4.70 ± 0.42	5.04 ± 0.72	5.92 ± 1.76	5.14 ± 1.19	4.96 ± 1.07
RBC (×1012 g/L)	4.76 ± 0.29	4.48 ± 0.40	4.83 ± 0.37	4.09 ± 0.16	3.95 ± 0.19	4.14 ± 0.32
Hb (g/L)	155.80 ± 8.93	151.20 ± 9.15	160.20 ± 8.44	133.40 ± 6.35	126.80 ± 8.58	134.00 ± 11.77
Hct (%)	45.46 ± 1.40	43.02 ± 2.53	45.80 ± 2.08	38.46 ± 1.70	37.02 ± 1.77	38.96 ± 3.06
MCV (Fl)	95.60 ± 3.08	95.20 ± 3.30	95.06 ± 3.59	94.02 ± 3.62	93.68 ± 2.95	94.08 ± 3.06
CK (U/L)	183.60 ± 48.44	451.80 ± 95.70 *	228.00 ± 48.84 #	80.80 ± 14.51	230.00 ± 29.80 *	98.40 ± 32.19 #
BU (mmol/L)	5.10 ± 0.43	6.25 ± 0.30 *	5.58 ± 0.83	4.87 ± 1.89	6.78 ± 1.66	5.27 ± 1.52
T (ng/dL)	666.60 ± 101.9	554.40 ± 135.07	580.40 ± 146.26	50.35 ± 12.94	34.17 ± 12.27 *	36.26 ± 15.35

* $p < 0.05$ vs. Pre-training; # $p < 0.05$ vs. Peak Training.

3.2. Results of Serum Metabolomics Detection

Assignment and Analyses of ^{1}H-NMR Spectra

The main compounds in the serum sample spectra of athletes were assigned by analyzing data such as chemical shifts, coupling constants and peak patterns, combined with public databases including the Human Metabolome Database (HMDB, http://www.hmdb.ca/) and the Biological Magnetic Resonance Data Bank (BMRB, https://bmrb.io), as well as relevant literature reports. A total of 22 metabolites were identified (Figure 2, Table 3), mainly including amino acids, carbohydrates, and nitrogen-containing compounds.

Table 3. Assignment results of serum metabolites in athletes via ^{1}H-NMR spectra.

No.	Compound Name	Chemical Shift (δ)	Peak Shape
1	Lipids	0.89	m
2	Leucine	0.95	d
3	Isoleucine	0.99	d
4	Valine	1.04	d

Table 3. *Cont.*

No.	Compound Name	Chemical Shift (δ)	Peak Shape
5	β-Hydroxybutyrate	1.15	dd
6	Lactic Acid	1.32	d
7	Alanine	1.48	d
8	N-Acetylglycoproteins	2.04	s
9	Acetoacetic Acid	2.27	s
10	Pyruvic Acid	2.37	s
11	Citric Acid	2.65	d
12	Creatine	3.03	s
13	Creatinine	3.04	t
14	Choline	3.20	s
15	Trimethylamine N-Oxide	3.23	m
16	Taurine	3.26	t
17	Proline	3.41	d
18	α-Glucose	3.47	d
19	Glycerol	3.53	s
20	β-Glucose	3.71	d
21	Glutamic Acid	3.72	t
22	Hippurate	3.98	d

s: single peak, d: double peak, dd: two double peaks, t: triple peak, m: multiple peaks.

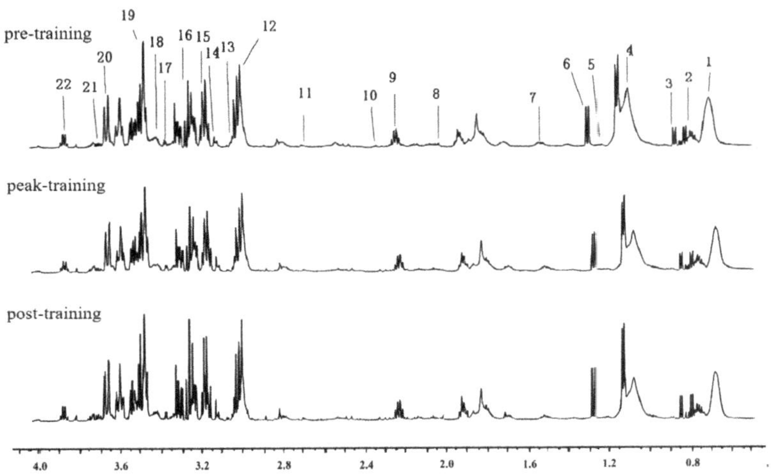

Figure 2. ^{1}H-NMR spectra of serum metabolites in athletes at different training stages.

3.3. *Formatting of Mathematical Components*

Results of Multivariate Statistical Analyses

Further model validation analyses using PLS-DA on serum samples from 10 freestyle wrestlers at different training stages resulted in the score plot shown in Figure 3. It is evident that the three groups of serum samples, namely pre-training, peak training, and post-training, are clearly separated, while samples within the same group achieved clustering.

To further identify metabolite differences between male wrestlers and female wrestlers across distinct training phases, athletes were categorized by sex. Metabolites were comparatively analyzed for male and female wrestlers before training, during peak training, and after training, resulting in OPLS-DA score plots and loading plots. For compounds with VIP $\geq$ 1 in the loading plot, one-way ANOVA was performed on their corresponding metabolite peak areas using SPSS software. This ultimately identified the significantly different metabolites ($p < 0.05$).

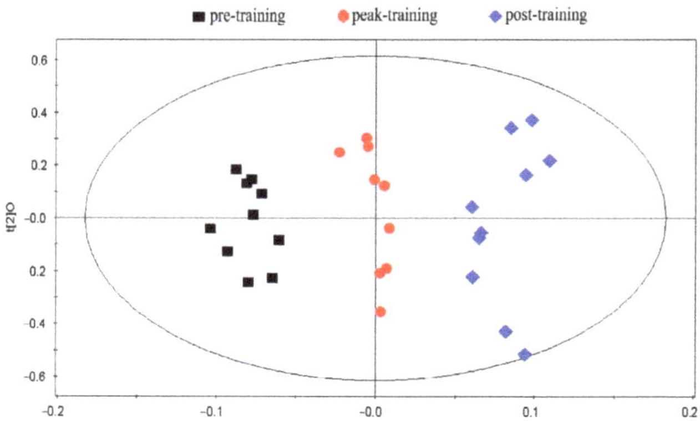

Figure 3. Score plot of PLS-DA analyses.

Compared to pre-training, a total of six differential metabolites were identified in the serum of male wrestlers in the peak training (Figure 4) (Table 4). Among these, the concentrations of leucine, isoleucine, valine, and lactate showed significant increases, while the concentrations of β-hydroxybutyrate and hippurate showed significant decreases. After adequate rest, the concentrations of leucine, isoleucine, valine, and lactate in male wrestlers decreased significantly compared to the peak training.

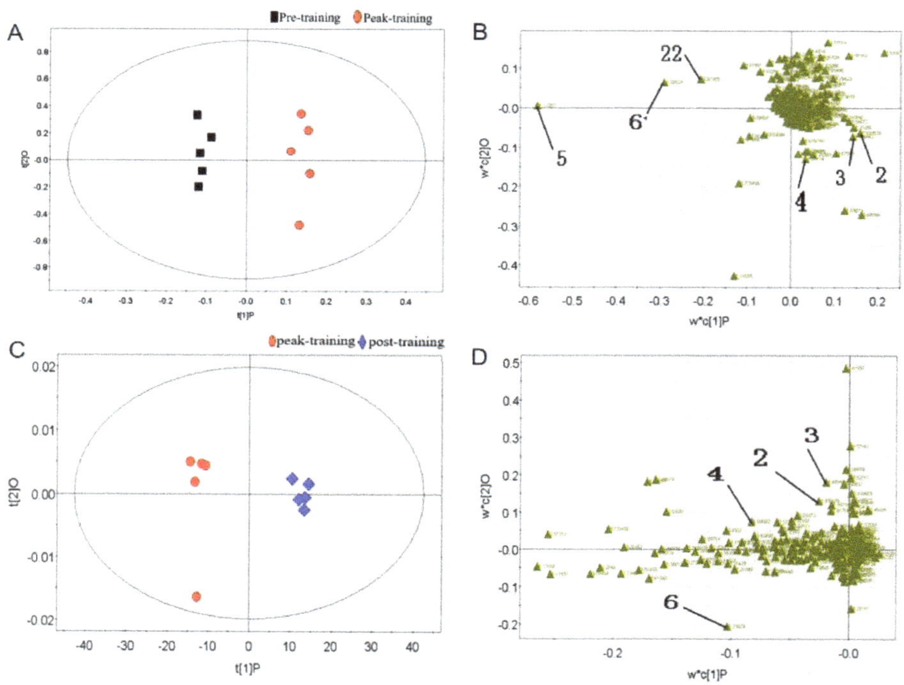

Figure 4. Multivariate data analyses of male wrestlers. (**A**) OPLS-DA score plot from pre-training and peak training; (**B**) load plot from pre-training and peak training; (**C**) OPLS-DA score plot from peak training and post-training; (**D**) load plot from peak training and post-training. 2: Leucine, 3: Isoleucine, 4: Valine, 5: β-hydroxybutyric acid, 6: Lactic acid, 22: Hippurate.

Table 4. Results of differentially metabolites in serum of male wrestlers.

No.	Metabolites	Chemical Shift (δ)	Mean ± SD	Peak-Training vs. Pre-Training	Post-Training vs. Peak-Training
1	Leucine	0.95 (d)	0.0009 ± 0.00054	↑	↓
2	Isoleucine	0.99 (d)	0.0009 ± 0.00050	↑	↓
3	Valine	1.05 (d)	0.0030 ± 0.00052	↑	↓
4	β-hydroxybutyric acid	1.17 (dd)	0.0698 ± 0.00989	↓	—
5	Lactic acid	1.32 (d)	0.0116 ± 0.0278	↑	↓
6	Hippurate	3.98 (d)	0.0060 ± 0.00112	↓	—

d: double peaks, dd: two double peaks; ↑: increase, ↓: decrease, —: no significant change.

Compared to pre-training, a total of three differential metabolites were identified in the serum of female wrestlers during peak training (Figure 5) (Table 5). Among these, pyruvate concentrations showed a significant increase, while valine concentrations exhibited a significant decrease ($p < 0.05$). After adequate rest, the concentrations of pyruvate decreased significantly, while those of creatine and valine increased significantly ($p < 0.05$) compared to the peak training.

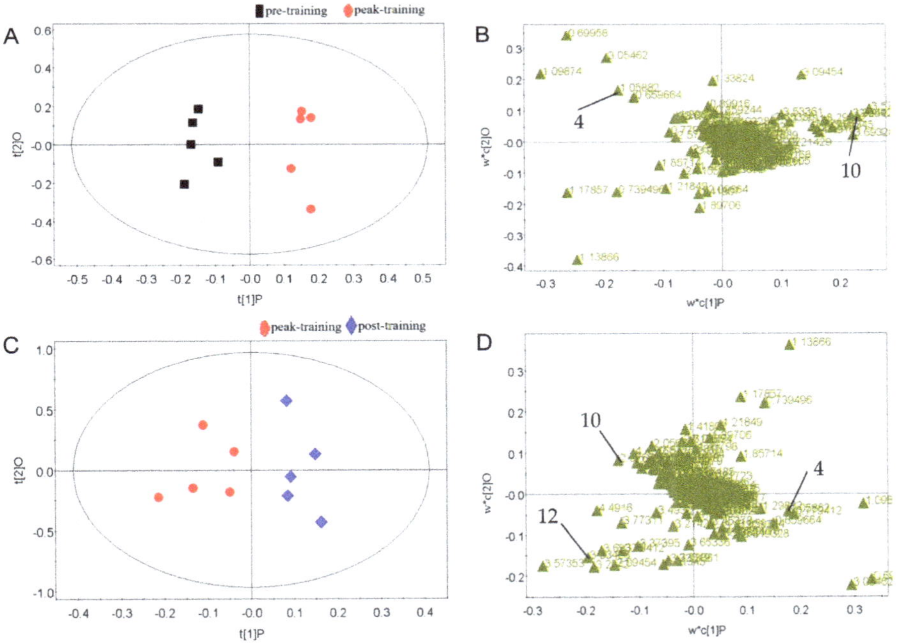

Figure 5. Multivariate data analyses of female wrestlers. (**A**) OPLS-DA score plot from pre-training and peak training; (**B**) load plot from pre-training and peak training; (**C**) OPLS-DA score plot from peak training and post-training; (**D**) load plot from peak training and post-training. 4: Valine, 10: Pyruvate, 12: Creatine.

Table 5. Results of differentially metabolites in serum of female wrestlers.

No.	Metabolites	Chemical Shift (δ)	Mean ± SD	Peak- Training vs. Pre-Training	Post-Training vs. Peak Training
1	Valine	1.05 (d)	0.0082 ± 0.00123	↓	↑
2	Pyruvate	2.37 (s)	0.0016 ± 0.00040	↑	↓
3	Creatine	3.03 (s)	0.0389 ± 0.00326	—	↑

d: double peaks, s: single-peak; ↑: increase, ↓: decrease, —: no significant change.

To further identify the metabolic pathways affected by the potential serum biomarkers in freestyle wrestlers, all screened potential biomarkers were imported into the online analysis platform MetaboAnalyst 5.0 (http://www.metaboanalyst.ca/), and a pathway impact plot was generated (Figure 6). The key metabolic pathways involved were derived from this plot. As shown in Table 6, freestyle wrestling training primarily affected four metabolic pathways: pyruvate metabolism, glycolysis or gluconeogenesis, citrate cycle (TCA cycle), and arginine and proline metabolism.

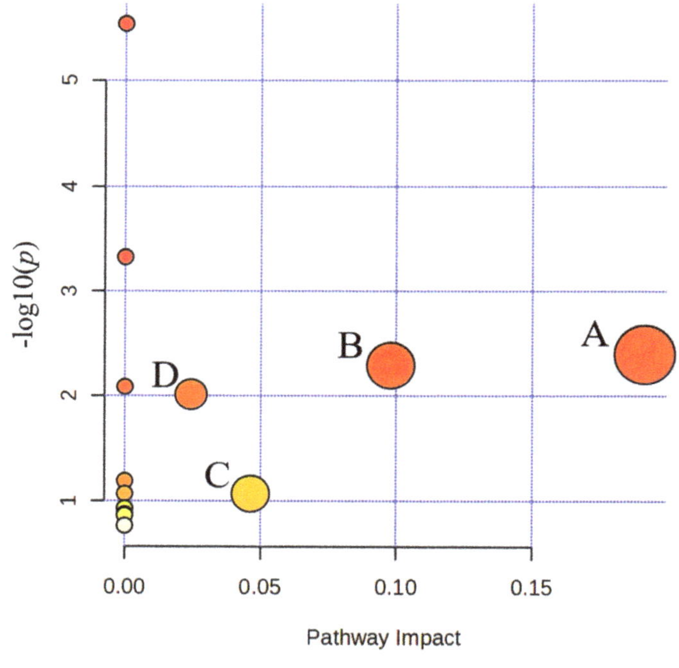

Figure 6. Summary diagram of pathway analyses with Met-PA. (A) Pyruvate metabolism; (B) Glycolysis or Gluconeogenesis; (C) Citrate cycle (TCA cycle); (D) Arginine and proline metabolism.

Table 6. Pathway enrichment analyses results from MetaboAnalyst 5.0.

Metabolic Pathway	Total	Hits	p	$-\log(p)$	Holm p	FDR	Impact
Pyruvate metabolism	23	2	0.004014	2.3964	0.31309	0.10248	0.19137
Glycolysis or Gluconeogenesis	26	2	0.0051238	2.2904	0.39453	0.10248	0.09785
Citrate cycle (TCA cycle)	20	1	0.084848	1.0714	1.0	0.7542	0.04634
Arginine and proline metabolism	36	1	0.0097248	2.0121	0.72936	0.12966	0.02442

4. Discussion

This study analyzed serum samples collected from freestyle wrestlers during different training phases of the intensive training cycle. By integrating conventional biochemical indicators with metabolomics technology, it explored the metabolic response characteristics and recovery status of athletes' bodies to training loads.

4.1. Training Load and Recovery Status Reflected by Conventional Biochemical Indicators

Conventional biochemical indicators serve as a critical window for evaluating exercise load and bodily adaptability [15]. This study found that both male wrestlers and female wrestlers showed a significant increase in the concentration of CK at the peak training compared to pre-training. CK serves as a sensitive indicator reflecting the extent of muscle damage and the intensity of exercise load [16]. The significant elevation of CK in athletes indicates that the training during this stage exerted substantial mechanical stimulation on their skeletal muscles, achieving the expected effect of targeted training intensity [17]. Notably, after an appropriate rest period, CK concentrations decreased significantly, suggesting that the athletes possessed a favorable short-term recovery capacity.

The changes in the concentration of BU mainly reflect the state of protein catabolism caused by exercise consumption and the body's potential for recovery [18]. Normally, the concentration of BU changes little during short-duration exercise (<30 min), with significant increases primarily observed after prolonged or high-intensity exercise [19]. It was found that an increase in the concentration of BU exceeding 3 mmol/L before and after training indicates excessive training volume leading to fatigue accumulation. An increase of approximately 2 mmol/L suggests a moderate training volume, while an increase of about 1 mmol/L indicates relatively insufficient training volume [20]. In this study, the concentration of BU in male wrestlers at peak training increased by only 1.15 mmol/L compared to pre-training, indicating that the training volume was relatively low. It suggests that the training load in the current phase is insufficient to effectively stimulate the bodies of male wrestlers, failing to fully activate deep physiological adaptation mechanisms.

Furthermore, the T concentration of female wrestlers showed a significant decreasing trend at the training peak. Testosterone is not only a key hormone reflecting anabolic status and bodily recovery capacity, but also plays a crucial role in enhancing the excitability of the central nervous system [21,22]. It can activate the protein synthesis enzyme system, and stimulate muscle cell receptors to promote muscle repair and growth [23]. Some studies showed that the serum concentration of T decreased by more than 25% after exercise and failed to recover after training, suggesting that athletes were not adapting to the training load and might be experiencing excessive fatigue, which required adjustments to the training schedule [24]. This study found that female wrestlers showed a significant decrease in the concentration of T during the peak training period compared to pre-training, exceeding the warning threshold. This indicates that the training load during this phase stimulated the athletes' bodies to a deeper degree, potentially approaching the upper limit of their capacity. Notably, after adequate rest, there was a significant rebound in the concentration of T among the female wrestlers. This suggests that although the load was substantial, it remained within the athletes' controllable range, and their bodies possessed a certain potential for recovery.

4.2. Sex-Specific Metabolic Response Patterns Revealed by Metabolomics

Metabolomics analyses further revealed the complex molecular-level responses of athletes to training stress and demonstrated significant sex differences.

4.2.1. Metabolic Response Characteristics of Male Wrestlers

A total of 6 serum metabolic biomarkers were identified to significantly change across the different training phases in male wrestlers: leucine, isoleucine, valine, β-hydroxybutyrate, lactate, and hippuric acid. Compared to pre-exercise, the concentrations of leucine, isoleucine, and valine (BCAAs) and lactate significantly increased at peak training, while the concentrations of β-hydroxybutyrate and hippuric acid significantly decreased. Compared to the peak training period, the concentrations of leucine, isoleucine, valine, and lactate decreased significantly post-exercise; there was no significant change in the concentrations of β-hydroxybutyrate and hippuric acid.

BCAAs, as essential amino acids, cannot be synthesized by the human body and must be obtained through dietary intake. During prolonged or high-intensity exercise, they undergo extensive catabolic metabolism in muscle tissue to provide energy [25]. Leucine metabolism can generate ketone bodies, isoleucine generates ketone bodies and acyl-CoA, while valine can ultimately generate glucose [26]. This study found that male wrestlers exhibited elevated BCAA concentrations during peak-training, contrasting with the declining BCAA trend observed in endurance-based prolonged exercise [27]. This discrepancy may be attributed to the intermittent, explosive nature of wrestling, which drives protein breakdown rates beyond oxidative utilization [28]. This finding is also consistent with the transient elevation of BCAA concentrations during anaerobic sprinting, suggesting sport-specific metabolic characteristics across different athletic disciplines [29]. Notably, the occurrence of exercise-induced fatigue is closely related to BCAA metabolism [30]. Prolonged high-load exercise can alter the plasma amino acid profile, increasing the ratio of AAAs to BCAAs [31]. This imbalance is believed to accelerate the transport of amino acids (particularly tryptophan) into the brain, promote serotonin synthesis, enhance central inhibition, and thereby induce central fatigue [32]. Therefore, the elevated serum BCAA concentrations observed at peak training in this study reflect their mobilization as energy substrates.

Lactic acid is the end product of glycolysis [33]; its significant increase at the peak-training directly confirms that anaerobic glycolysis is one of the primary energy supply pathways for male wrestlers during wrestling training. Hippurate is synthesized in the liver from benzoate, which is largely derived from gut microbiota metabolism [34,35]. In this study, hippurate significantly decreased at the training peak and did not rebound significantly during recovery. This decrease at the peak-training may be related to exercise-induced alterations in gut microbial composition or activity [36]. The lack of significant rebound during post-training may indicate the long-term effects of training load on this gut–liver metabolic axis [37].

4.2.2. Metabolic Response Characteristics of Female Wrestlers

A total of three significantly different metabolic markers were identified in female wrestlers across different training phases: valine, pyruvate, and creatine, with fewer markers identified compared to male wrestlers. Compared to pre-training, there was a significant decrease in the concentration of valine and a significant increase in the concentration of pyruvate at the peak training, while the concentration of creatine showed no significant change. Compared to peak training, the concentration of valine and creatine significantly increased post-training, while the concentration of pyruvate significantly decreased.

Valine is an important glucogenic amino acid, and its metabolism ultimately generates glucose [38]. Unlike the increasing trend observed in male wrestlers, female wrestlers showed a significant decrease in the concentration of valine in serum during peak training. This reveals sex differences in energy metabolism. The decline in the valine concentrations of female wrestlers indicates its rapid uptake and breakdown for energy to meet the demands of high-intensity training. This finding correlates with the significantly reduced the concentration of T observed in female wrestlers in this study, suggesting the body is in a more pronounced catabolic state during this phase [39]. Research indicates that women often rely more heavily on fat oxidation during exercise [40]. However, during high-intensity interval training (HIIT) with substantial glycogen depletion, they may instead enhance amino acid gluconeogenesis, leading to increased consumption of valine [41]. This suggests the need to focus on the intake and supplementation of BCAAs for female wrestlers to support training and delay fatigue. Pyruvic acid is a key intermediate product of glucose metabolism [42]. The significantly increased concentration of pyruvate at peak training reflects the overall activation of energy metabolism pathways to meet training demands [43].

Unlike male wrestlers, this study did not observe a significant increase in the concentration of lactate at peak training in female wrestlers. Since lactic acid is converted from pyruvic acid under hypoxic conditions [44], this finding indicates that during the training phase, female wrestlers show relatively lower activity in the glycolytic energy pathway compared to male wrestlers, or in other words, a smaller proportion of pyruvate flows into the lactate pathway. Combined with the significant decrease in the concentration of valine, this further supports that catabolism plays a substantial role in energy supply for female wrestlers during this phase. There was no significant change in the concentration of creatine at peak training intensity, but it increased significantly after training. Creatine participates in the phosphocreatine system and energy buffering [45]. Female wrestlers showed significantly increased serum creatine concentrations post-training, potentially due to active uptake and resynthesis of creatine by muscle cells to prepare for subsequent energy demands. Alternatively, this may reflect an overall improvement in the anabolic environment, promoting endogenous creatine synthesis [46].

4.2.3. Pathway Perturbations and Training Adaptation

The serum metabolomic profiles effectively capture the comprehensive physiological demands of freestyle wrestling training across different phases. The significantly elevated concentrations of lactate and pyruvate, particularly in male wrestlers, directly confirm the activation of anaerobic glycolysis as a primary energy pathway during high-intensity efforts [47,48].

The alterations in branched-chain amino acids (BCAAs) reveal distinct sex-specific metabolic responses. The increased BCAA concentrations in male wrestlers suggest a sport-specific pattern where the rate of protein catabolism and BCAA mobilization exceeds their oxidative utilization [49]. In contrast, the decreased valine concentration in female wrestlers indicates enhanced utilization of glucogenic amino acids for energy production, highlighting fundamental differences in substrate utilization between sexes [50].

The observed changes in other metabolites provide additional insights into metabolic adaptation. The decreased β-hydroxybutyrate concentration during peak training suggests a redirection of acetyl-CoA from ketogenesis toward the TCA cycle to meet energy demands [51]. Meanwhile, the significantly increased creatine concentration in female wrestlers during recovery indicates active replenishment of the phosphocreatine system, reflecting an anabolic response following training stress [52].

4.3. Limitations of the Study

This study has several limitations that should be considered when interpreting the results. First, the relatively small sample size and the exclusive inclusion of elite-level wrestlers may limit the generalizability of the findings to athletes of different competitive levels or training backgrounds. However, a post hoc power analysis was conducted based on the observed large effect sizes in biochemical markers (e.g., Creatine Kinase, Cohen's d > 3.0) and the clear separation in multivariate models. The analysis indicated that for detecting changes of such magnitude, the sample size provided a statistical power (1-β) exceeding 0.99, which is well above the conventional threshold of 0.80. This suggests that the study was sufficiently powered to identify the primary, large-effect metabolic and physiological responses reported. Nevertheless, the small sample size remains a constraint for detecting subtle effects, ensuring robust generalizability, and for conducting more complex subgroup analyses.

Second, although ^{1}H-NMR metabolomics provides high reproducibility and is suitable for untargeted profiling, its comparatively lower sensitivity than mass spectrometry-based approaches may have led to the under-detection of certain low-abundance metabolites that could be relevant to exercise-induced metabolic adaptations.

Third, despite efforts to standardize the participants' diet and living conditions, potential confounding factors—such as individual nutritional intake, hydration status, and, for female athletes, menstrual cycle phase—were not strictly monitored or controlled. These factors may have introduced additional variability into the metabolic data.

Finally, the metabolic profiling was performed over a single training microcycle. Although this approach offers valuable insights into short-term dynamics, it does not reflect long-term adaptive responses or potential metabolic periodization across a full training macrocycle. Future longitudinal studies involving larger cohorts, integrated multi-omics methodologies, and detailed workload monitoring are warranted to validate and extend these preliminary findings.

5. Conclusions

Serum differential metabolites in freestyle wrestlers across different training phases were primarily enriched in four metabolic pathways: pyruvate metabolism, glycolysis or gluconeogenesis, citrate cycle (TCA cycle), and arginine and proline metabolism. These pathways predominantly regulate energy metabolism and amino acid metabolism, suggesting metabolic adaptations under exercise load. This indicates that relevant metabolic biomarkers may serve as new targets for assessing athletes' metabolic status and the extent of recovery.

This study combines traditional biochemical monitoring with metabolomics analyses to provide a multiple-aspect perspective on the physiological functions and metabolic characteristics of freestyle wrestlers. The information of small-molecule metabolites detected by ^{1}H NMR metabolomics reflects the training load and physiological status of athletes of different sexes, providing a basis for training monitoring and scientific training. Future studies may explore larger sample sizes and multimodal monitoring to further illuminate the relationship between metabolites, exercise fatigue, and training load, thereby enabling a more comprehensive assessment of wrestling's metabolic characteristics.

Author Contributions: Conceptualization: Y.H.; resources: Y.H. and W.Z.; methodology: X.L. (Xiaonan Li) and X.L. (Xiangyu Liu); investigation: J.L. and X.L. (Xiaonan Li); data curation: X.L. (Xiangyu Liu); formal analyses: X.L. (Xiaonan Li); Funding acquisition: Y.H.; writing—original draft preparation: J.L.; writing—review and editing: X.L. (Xiangyu Liu) and X.L. (Xiaonan Li); supervision: Y.L., W.Z. and Y.H. All authors have read and agreed to the published version of the manuscript.

Funding: This work was supported by the Shanxi Provincial Key R&D Program Project (Grant No. 202103021224027), the Key Educational and Teaching Reform Research Project of Huainan Normal University (Grant No. 2024hsjxtd04).

Institutional Review Board Statement: This study was conducted in accordance with the Declaration of Helsinki and approved by the Research Ethics Committee of the University of Shanxi (Approval Code: SXULL2024110, approved on 15 November 2024).

Informed Consent Statement: Informed consent was obtained from all participants involved in the study.

Data Availability Statement: The original contributions presented in this study are included in the article. Further inquiries can be directed to the corresponding authors.

Conflicts of Interest: All authors declare that they have no competing interests.

References

1. He, Z.-H.; Feng, L.-S.; Zhang, H.-J.; Xu, K.-Y.; Chi, F.-T.; Tao, D.-L.; Liu, M.-Y.; Lucia, A.; Fleck, S.J. Physiological profile of elite Chinese female wrestlers. *J. Strength Cond. Res.* **2013**, *27*, 2374–2395. [CrossRef] [PubMed]
2. Chaabene, H.; Negra, Y.; Bouguezzi, R.; Mkaouer, B.; Franchini, E.; Julio, U.; Hachana, Y. Physical and Physiological Attributes of Wrestlers: An Update. *J. Strength Cond. Res.* **2017**, *31*, 1411–1442. [CrossRef] [PubMed]
3. Franchini, E.; Brito, C.J.; Artioli, G.G. Weight loss in combat sports: Physiological, psychological and performance effects. *J. Int. Soc. Sports Nutr.* **2012**, *9*, 52. [CrossRef]
4. Minetti, A.E.; Ardigò, L.P.; Saibene, F. Mechanical determinants of gradient walking energetics in man. *J. Physiol.* **1993**, *472*, 725–735. [CrossRef]
5. Fogarasi, A.; Gonzalez, K.; Dalamaga, M.; Magkos, F. The Impact of the Rate of Weight Loss on Body Composition and Metabolism. *Curr. Obes. Rep.* **2022**, *11*, 33–44. [CrossRef]
6. Hartmann, U.; Mester, J. Training and overtraining markers in selected sport events. *Med. Sci. Sports Exerc.* **2000**, *32*, 209–215. [CrossRef]
7. Lee, E.C.; Fragala, M.S.; Kavouras, S.A.; Queen, R.M.; Pryor, J.L.; Casa, D.J. Biomarkers in Sports and Exercise: Tracking Health, Performance, and Recovery in Athletes. *J. Strength Cond. Res.* **2017**, *31*, 2920–2937. [CrossRef]
8. Patti, G.J.; Yanes, O.; Siuzdak, G. Innovation: Metabolomics: The apogee of the omics trilogy. *Nat. Rev. Mol. Cell Biol.* **2012**, *13*, 263–269. [CrossRef]
9. Wishart, D.S. Metabolomics for Investigating Physiological and Pathophysiological Processes. *Physiol. Rev.* **2019**, *99*, 1819–1875. [CrossRef]
10. Khoramipour, K.; Sandbakk, O.; Keshteli, A.H.; Gaeini, A.A.; Wishart, D.S.; Chamari, K. Metabolomics in Exercise and Sports: A Systematic Review. *Sports Med.* **2022**, *52*, 547–583. [CrossRef] [PubMed]
11. Belhaj, M.R.; Lawler, N.G.; Hoffman, N.J. Metabolomics and Lipidomics: Expanding the Molecular Landscape of Exercise Biology. *Metabolites* **2021**, *11*, 151. [CrossRef]
12. Heaney, L.M.; Deighton, K.; Suzuki, T. Non-targeted metabolomics in sport and exercise science. *J. Sports Sci.* **2019**, *37*, 959–967. [CrossRef] [PubMed]
13. MoTrPAC Study Group. Temporal dynamics of the multi-omic response to endurance exercise training. *Nature* **2024**, *629*, 174–183. [CrossRef] [PubMed]
14. Puigarnau, S.; Fernàndez, A.; Obis, E.; Jové, M.; Castañer, M.; Pamplona, R.; Portero-Otin, M.; Camerino, O. Metabolomics reveals that fittest trail runners show a better adaptation of bioenergetic pathways. *J. Sci. Med. Sport* **2022**, *25*, 425–431. [CrossRef] [PubMed]
15. Robbins, J.M.; Peterson, B.; Schranner, D.; Tahir, U.A.; Rienmuller, T.; Deng, S.; Keyes, M.J.; Katz, D.H.; Beltran, P.M.J.; Barber, J.L.; et al. Human plasma proteomic profiles indicative of cardiorespiratory fitness. *Nat. Metab.* **2021**, *3*, 786–797. [CrossRef]
16. Saidi, K.; Abderrahman, A.B.; Hackney, A.C.; Bideau, B.; Zouita, S.; Granacher, U.; Zouhal, H. Hematology, Hormones, Inflammation, and Muscle Damage in Elite and Professional Soccer Players: A Systematic Review with Implications for Exercise. *Sports Med.* **2021**, *51*, 2607–2627. [CrossRef]
17. Moore, E.; Fuller, J.T.; Buckley, J.D.; Saunders, S.; Halson, S.L.; Broatch, J.R.; Bellenger, C.R. Impact of Cold-Water Immersion Compared with Passive Recovery Following a Single Bout of Strenuous Exercise on Athletic Performance in Physically Active Participants: A Systematic Review with Meta-analysis and Meta-regression. *Sports Med.* **2022**, *52*, 1667–1688. [CrossRef]
18. Urhausen, A.; Gabriel, H.; Kindermann, W. Blood hormones as markers of training stress and overtraining. *Sports Med.* **1995**, *20*, 251–276. [CrossRef]
19. Bessa, A.; Nissenbaum, M.; Monteiro, A.; Gandra, P.G.; Nunes, L.S.; Bassini-Cameron, A.; Werneck-de-Castro, J.P.; de Macedo, D.V.; Cameron, L.C. High-intensity ultraendurance promotes early release of muscle injury markers. *Br. J. Sports Med.* **2008**, *42*, 889–893. [CrossRef]
20. Emambokus, N.; Granger, A.; Messmer-Blust, A. Exercise Metabolism. *Cell Metab.* **2015**, *22*, 1. [CrossRef]

21. Wu, L.; Qu, J.; Mou, L.; Liu, C. Apigenin improves testosterone synthesis by regulating endoplasmic reticulum stress. *Biomed. Pharmacother.* **2024**, *177*, 117075. [CrossRef]
22. Kelava, I.; Chiaradia, I.; Pellegrini, L.; Kalinka, A.T.; Lancaster, M.A. Androgens increase excitatory neurogenic potential in human brain organoids. *Nature* **2022**, *602*, 112–116. [CrossRef] [PubMed]
23. Oura, M.; Son, B.K.; Song, Z.; Toyoshima, K.; Nanao-Hamai, M.; Ogawa, S.; Akishita, M. Testosterone/androgen receptor antagonizes immobility-induced muscle atrophy through Inhibition of myostatin transcription and inflammation in mice. *Sci. Rep.* **2025**, *15*, 10568. [CrossRef] [PubMed]
24. Rogerson, D.; Nolan, D.; Androulakis Korakakis, P.; Immonen, V.; Wolf, M.; Bell, L. Deloading Practices in Strength and Physique Sports: A Cross-sectional Survey. *Sports Med. Open* **2024**, *10*, 26. [CrossRef] [PubMed]
25. Li, G.; Li, Z.; Liu, J. Amino acids regulating skeletal muscle metabolism: Mechanisms of action, physical training dosage recommendations and adverse effects. *Nutr. Metab.* **2024**, *21*, 41. [CrossRef]
26. Adeva-Andany, M.M.; Lopez-Maside, L.; Donapetry-Garcia, C.; Fernandez-Fernandez, C.; Sixto-Leal, C. Enzymes involved in branched-chain amino acid metabolism in humans. *Amino Acids* **2017**, *49*, 1005–1028. [CrossRef]
27. Gibala, M.J. Protein Metabolism and Endurance Exercise. *Sports Med.* **2007**, *37*, 337–340. [CrossRef]
28. Shimomura, Y.; Honda, T.; Shiraki, M.; Murakami, T.; Sato, J.; Kobayashi, H.; Mawatari, K.; Obayashi, M.; Harris, R.A. Branched-chain amino acid catabolism in exercise and liver disease. *J. Nutr.* **2006**, *136*, 250s–253s. [CrossRef]
29. Gawedzka, A.; Grandys, M.; Duda, K.; Zapart-Bukowska, J.; Zoladz, J.A.; Majerczak, J. Plasma BCAA concentrations during exercise of varied intensities in young healthy men-the impact of endurance training. *PeerJ* **2020**, *8*, e10491. [CrossRef]
30. Luan, C.; Wang, Y.; Li, J.; Zhou, N.; Song, G.; Ni, Z.; Xu, C.; Tang, C.; Fu, P.; Wang, X.; et al. Branched-Chain Amino Acid Supplementation Enhances Substrate Metabolism, Exercise Efficiency and Reduces Post-Exercise Fatigue in Active Young Males *Nutrients* **2025**, *17*, 1290. [CrossRef]
31. Xu, M.; Hu, D.; Liu, X.; Li, Z.; Lu, L. Branched-Chain Amino Acids and Inflammation Management in Endurance Sports: Molecular Mechanisms and Practical Implications. *Nutrients* **2025**, *17*, 1335. [CrossRef]
32. Davis, J.M.; Alderson, N.L.; Welsh, R.S. Serotonin and central nervous system fatigue: Nutritional considerations. *Am. J. Clin. Nutr.* **2000**, *72*, 573s–578s. [CrossRef] [PubMed]
33. Brooks, G.A. The Science and Translation of Lactate Shuttle Theory. *Cell Metab.* **2018**, *27*, 757–785. [CrossRef] [PubMed]
34. Williams, H.R.; Cox, I.J.; Walker, D.G.; Cobbold, J.F.; Taylor-Robinson, S.D.; Marshall, S.E.; Orchard, T.R. Differences in gut microbial metabolism are responsible for reduced hippurate synthesis in Crohn's disease. *BMC Gastroenterol.* **2010**, *10*, 108 [CrossRef]
35. De Simone, G.; Balducci, C.; Forloni, G.; Pastorelli, R.; Brunelli, L. Hippuric acid: Could became a barometer for frailty and geriatric syndromes? *Ageing Res. Rev.* **2021**, *72*, 101466. [CrossRef]
36. Pallister, T.; Jackson, M.A.; Martin, T.C.; Zierer, J.; Jennings, A.; Mohney, R.P.; MacGregor, A.; Steves, C.J.; Cassidy, A.; Spector, T.D.; et al. Hippurate as a metabolomic marker of gut microbiome diversity: Modulation by diet and relationship to metabolic syndrome. *Sci. Rep.* **2017**, *7*, 13670. [CrossRef]
37. E, H.-W.; Kosmol, A.; Gajewski, J.J.B.o.S. Aerobic fitness of elite female and male wrestlers. *Biol. Sport* **2009**, *26*, 339–348. [CrossRef]
38. Mansoori, S.; Ho, M.Y.; Ng, K.K.; Cheng, K.K. Branched-chain amino acid metabolism: Pathophysiological mechanism and therapeutic intervention in metabolic diseases. *Obes. Rev. Off. J. Int. Assoc. Study Obes.* **2025**, *26*, e13856. [CrossRef]
39. Carter, S.L.; Rennie, C.; Tarnopolsky, M.A. Substrate utilization during endurance exercise in men and women after endurance training. *Am. J. Physiol. Endocrinol. Metab.* **2001**, *280*, E898–E907. [CrossRef]
40. Sanchez, B.N.; Volek, J.S.; Kraemer, W.J.; Saenz, C.; Maresh, C.M. Sex Differences in Energy Metabolism: A Female-Oriented Discussion. *Sports Med.* **2024**, *54*, 2033–2057. [CrossRef]
41. Salem, A.; Ben Maaoui, K.; Jahrami, H.; AlMarzooqi, M.A.; Boukhris, O.; Messai, B.; Clark, C.C.T.; Glenn, J.M.; Ghazzaoui, H.A.; Bragazzi, N.L.; et al. Attenuating Muscle Damage Biomarkers and Muscle Soreness After an Exercise-Induced Muscle Damage with Branched-Chain Amino Acid (BCAA) Supplementation: A Systematic Review and Meta-analysis with Meta-regression *Sports Med. Open* **2024**, *10*, 42. [CrossRef]
42. Sousa, J.; Westhoff, P.; Methling, K.; Lalk, M. The Absence of Pyruvate Kinase Affects Glucose-Dependent Carbon Catabolite Repression in Bacillus subtilis. *Metabolites* **2019**, *9*, 216. [CrossRef]
43. Olek, R.A.; Kujach, S.; Radak, Z. Current knowledge about pyruvate supplementation: A brief review. *Sports Med. Health Sci.* **2024**, *6*, 295–301. [CrossRef] [PubMed]
44. Rabinowitz, J.D.; Enerback, S. Lactate: The ugly duckling of energy metabolism. *Nat. Metab.* **2020**, *2*, 566–571. [CrossRef] [PubMed]
45. Bonilla, D.A.; Kreider, R.B.; Stout, J.R.; Forero, D.A.; Kerksick, C.M.; Roberts, M.D.; Rawson, E.S. Metabolic Basis of Creatine in Health and Disease: A Bioinformatics-Assisted Review. *Nutrients* **2021**, *13*, 1238. [CrossRef] [PubMed]
46. Cooper, R.; Naclerio, F.; Allgrove, J.; Jimenez, A. Creatine supplementation with specific view to exercise/sports performance: An update. *J. Int. Soc. Sports Nutr.* **2012**, *9*, 33. [CrossRef]

7. Mirzaei, B.; Curby, D.G.; Rahmani-Nia, F.; Moghadasi, M. Physiological profile of elite Iranian junior freestyle wrestlers. *J. Strength Cond. Res.* **2009**, *23*, 2339–2344. [CrossRef]

8. San-Millan, I.; Stefanoni, D.; Martinez, J.L.; Hansen, K.C.; D'Alessandro, A.; Nemkov, T. Metabolomics of Endurance Capacity in World Tour Professional Cyclists. *Front. Physiol.* **2020**, *11*, 578. [CrossRef]

9. Lamont, L.S.; McCullough, A.J.; Kalhan, S.C. Gender differences in the regulation of amino acid metabolism. *J. Appl. Physiol.* **2003**, *95*, 1259–1265. [CrossRef]

0. Krumsiek, J.; Mittelstrass, K.; Do, K.T.; Stuckler, F.; Ried, J.; Adamski, J.; Peters, A.; Illig, T.; Kronenberg, F.; Friedrich, N.; et al. Gender-specific pathway differences in the human serum metabolome. *Metabol. Off. J. Metabol. Soc.* **2015**, *11*, 1815–1833. [CrossRef]

1. Evans, M.; Cogan, K.E.; Egan, B. Metabolism of ketone bodies during exercise and training: Physiological basis for exogenous supplementation. *J. Physiol.* **2017**, *595*, 2857–2871. [CrossRef]

2. Tam, R.; Mitchell, L.; Forsyth, A. Does Creatine Supplementation Enhance Performance in Active Females? A Systematic Review. *Nutrients* **2025**, *17*, 238. [CrossRef]

metabolites

MDPI

Article

Metabolomic Profiling of the Striatum in *Shank3* Knockout ASD Rats: Effects of Early Swimming Regulation

Yunchen Meng [1,*], Yiling Hu [1], Yaqi Xue [2] and Zhiping Zhen [2,*]

[1] Department of Physical Education and Research, China University of Mining and Technology—Beijing, Beijing 100083, China; zqt2410707005p@student.cumtb.edu.cn

[2] College of P.E and Sports, Beijing Normal University, Beijing 100875, China; xueyaqi555@mail.bnu.edu.cn

* Correspondence: yc_meng@cumtb.edu.cn (Y.M.); zzpxt@bnu.edu.cn (Z.Z.)

Abstract: Objectives: This study aimed to investigate the regulatory impact of early swimming intervention on striatal metabolism in *Shank3* gene knockout ASD model rats. **Methods:** *Shank3* gene knockout exon 11–21 male 8-day-old SD rats were used as experimental subjects and randomly divided into the following three groups: a *Shank3* knockout control group (KC), a wild-type control group (WC) from the same litter, and a *Shank3* knockout swimming group (KS). The rats in the exercise group received early swimming intervention for 8 weeks starting at 8 days old. LC-MS metabolism was employed to detect the changes in metabolites in the striatum. **Results:** There were 17 differential metabolites (14 down-regulated) between the KC and WC groups, 19 differential metabolites (18 up-regulated) between the KS and KC groups, and 22 differential metabolites (18 up-regulated) between the KS and WC groups. **Conclusions:** The metabolism of striatum in *Shank3* knockout ASD model rats is disrupted, involving metabolites related to synaptic morphology, and the Glu and GABAergic synapses are abnormal. Early swimming intervention regulated the striatal metabolome group of the ASD model rats, with differential metabolites primarily related to nerve development, synaptic membrane structure, and synaptic signal transduction.

Keywords: *Shank3*; early swimming; striatum; metabolism; neurotransmitter

1. Introduction

Autism spectrum disorder (ASD) is a complex neurodevelopmental condition marked by social impairments, repetitive and stereotyped behaviors, and a spectrum of symptoms encompassing cognitive deficits, motor dysfunction, intellectual disability, and attention deficit hyperactivity disorder (ADHD) [1,2]. The global incidence of ASD has risen significantly, posing a major public health challenge that affects children's well-being. According to a 2021 report by the Centers for Disease Control and Prevention (CDC), the prevalence of ASD among 8-year-olds was 1 in 44, representing a staggering 240% increase from 1 in 150 in 2000 [3]. In China, nearly 1% of children suffer from varying degrees of ASD, with the prevalence rate continuing to climb annually [4]. Given the high incidence and lifelong disability associated with ASD, early diagnosis and intervention are pivotal in its prevention and management [5].

The etiology of ASD is multifaceted, involving intricate interactions between genetic and environmental factors. While the precise causes remain unknown for at least 60% of ASD cases [6], extensive research has established that ASD originates in the early stages of brain development and persists throughout life. The genetic underpinnings of ASD are complex, characterized by widespread variations in genetic networks. Mutations or

 https://doi.org/10.3390/metabo1502013

deletions in numerous genes, notably the *Shank3* gene, have been implicated in the pathogenesis of ASD [7,8]. The protein encoded by the *Shank3* gene serves as a vital scaffolding component in the postsynaptic density (PSD) of excitatory synapses, playing a crucial role in synapse formation and stabilization and the regulation of synaptic transmission. Disruptions in the *Shank3* gene can impair normal synaptic function, leading to aberrant neuronal connections and, subsequently, triggering ASD symptoms [9]. Consequently, the genetic knockout of *Shank3* has emerged as a valuable animal model for studying autism.

In recent years, research into the pathophysiology of ASD has intensified, leading to a growing interest in exercise intervention as a non-pharmacological treatment option. Swimming, a whole-body exercise that combines motor stimulation with multi-sensory stimulation from a rich environment, is recognized as an effective means of promoting neurodevelopment and neural plasticity [10]. Appropriate swimming exercise not only bolsters cardiopulmonary function and enhances physical fitness but also exerts positive effects on the central nervous system, improving neurotransmitter release and fostering synaptic plasticity [11]. Studies on swimming interventions for ASD patients have documented its beneficial impact on ASD symptoms. Fragala-Pinkham et al. reported that a 14-week swimming program improved cardiopulmonary endurance and physical fitness in children with disabilities, including those with ASD [12]. Pan's research demonstrated that aquatic therapy significantly enhanced muscle strength, endurance, and social interaction skills in children with ASD [13]. Furthermore, Ennis et al. found that swimming interventions improved social, emotional, and learning outcomes in children with ASD [14]. These findings suggest that swimming exercise may ameliorate ASD symptoms by promoting neural plasticity and enhancing synaptic function. However, despite promising initial evidence, the specific mechanisms underlying the benefits of swimming intervention in ASD treatments remain incompletely understood. Particularly in the context of the *Shank3* gene knockout ASD animal model, further investigation is needed to determine whether swimming interventions can modulate striatal synaptic function to alleviate ASD symptoms. The striatum, a key component of the central nervous system, plays a crucial role in regulating various functions, including movement, cognition, and emotion. Structural and functional abnormalities in the striatum have been widely documented in individuals with ASD [15]. Therefore, exploring the effects of swimming interventions on striatal synaptic function in *Shank3* gene knockout rats with ASD is of paramount importance in elucidating the mechanisms underlying the therapeutic benefits of swimming.

In a previous study, *Shank3*$^{-/-}$ rats exhibited social impairments and stereotypic behaviors. Notably, these behavioral abnormalities showed improvement when the rats underwent early swimming exercise interventions [9]. A transcriptomic analysis of the striatum revealed differential gene expression related to synaptic structure and function between the *Shank3*$^{-/-}$ and *Shank3*$^{+/+}$ rats, further highlighting the differences between the *Shank3*$^{-/-}$ control rats and those subjected to swimming interventions [10]. The altered striatal transcriptome is likely to exert widespread influences on neurotransmitter release within the striatum. Abnormalities in neurotransmission, including glutamate, GABA, dopamine, serotonin, opioids, and oxytocin, have been reported in ASD patients [16,17].

Metabolites are small molecules produced as byproducts of enzymatic reactions and may serve as final or intermediate products. The abundance and composition of metabolites within a tissue directly reflect changes in the genome, transcriptome, and proteome [18]. Consequently, an emerging area of central nervous system (CNS) research has focused on the study and characterization of the metabolome. Among the metabolites relevant to CNS research are energy substrates, neurotransmitters, neurochemicals, and structural lipids. In this study, we employed a metabolomic approach to further investigate the effects of *Shank3* knockout and early swimming interventions on the metabolite profiles in rat striatum.

2. Materials and Methods

2.1. Experimental Animals and Grouping

Male specific-pathogen-free (SPF) Sprague Dawley (SD) *Shank3*$^{-/-}$ rats at the age of 8 days old were obtained from the Peking University School of Medicine. The germline deletion of *Shank3*, specifically targeting exon 11–21, was generated using CRISPR-Cas9 technology [19]. The *Shank3*$^{-/+}$ rats, both male and female, were generated by backcrossing *Shank3*$^{-/-}$ rats with *Shank3*$^{+/+}$ SD rats. Subsequently, the *Shank3*$^{-/+}$ rats were bred to obtain both SD *Shank3*$^{+/+}$ and *Shank3*$^{-/-}$ rats for the experimental groups. The rats were randomly allocated into the following three groups based on their genotypes and treatments: the *Shank3*$^{-/-}$ control group (KC; n = 3), *Shank3*$^{+/+}$ littermates (WC; n = 3), and the *Shank3*$^{-/-}$ swimming group (KS; n = 3).

The rats were housed in cages containing 3 animals each, with free access to food and water, in a room maintained at a temperature of 20–25 °C, with 45–55% humidity and a 12-h light/dark cycle. All animal procedures and care were conducted in strict accordance with the ethical standards and protocols outlined by the National Institutes of Health (US) [cite PMID 21595115] and were approved by the Biomedical Ethics Committee of Peking University (ethics number LA2021552). After the experiment, all the animals were euthanized by intraperitoneal injection of barbiturates.

2.2. Early Swimming Exercise Program

The swimming protocol employed in this study was adapted from a prior study by Muniz [20] and was performed on the rats in the KS group as depicted in Figure 1. An initial adaptation to swimming was implemented for the rats in the KS group, as illustrated in Figure 1 [10,21]. The initial acclimation to swimming took place during the post-natal days (PND) 8–10 in a cage measuring 485 × 350 × 200 mm that was filled with water maintained at a temperature of 32 ± 1 °C. The duration of each session was 2 min on PND 8, 5 min on PND 9, and 10 min on PND 10. On PND8, the water level was at the height of the rats' legs. On PND9, the water level was at the height of the rats' stomachs. On PND10, the water level was at the height of the rats' necks. Starting from PND 13, the swimming intervention was conducted in a circular tank (150 cm in diameter and 100 cm in height) filled with temperature-controlled water (32 ± 1 °C) to a depth of 50 cm, a setup that remained consistent for the remainder of the experiment. The duration of each session was 15 min on PND 13, 20 min on PND 14, and 25 min on PND 15. Between PND16–26, the rats engaged in 30 min of swimming per day, with 5 consecutive days of swimming followed by 2 days of rest per week. The duration of the swimming sessions was extended to 40 min per day from PND 27 to 60.

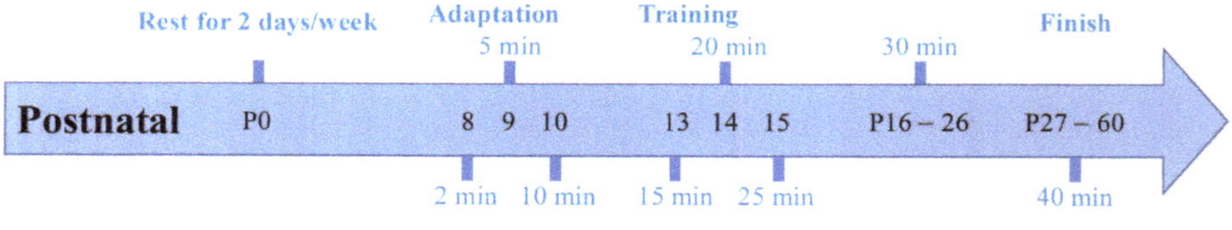

Figure 1. Early swimming intervention program [10]. The protocol included swimming from postnatal day (P) 8 to 60 and always at the same time (17:00 to 18:00), with rest periods for two consecutive rest days by week. The pups practiced swimming for 2, 5, and 10 min on P8, 9, and 10, respectively, and rested on P11 and 12. On P13, the rats swam for 15 and 20 min on P14 and 25 min on P15. From P16 to 26, the rats swam for 30 min/day, 5 times per week. From days 27 to 60, the rats swam for 40 min/day, 5 times per week.

2.3. *Striatal Metabolite Analysis*

2.3.1. Chemicals

This experiment utilized a variety of solutions and reagents. Liquid chromatography-mass-spectrometry-grade methanol and acetonitrile were both purchased from Thermo (Waltham, MA, USA). We sourced 2-chlorophenylalanine from Aladdin (Carlsbad, CA, USA). High-performance liquid-chromatography-grade formic acid was obtained from TCI (Tokyo, Japan). Liquid-chromatography-grade ammonium formate was purchased from Sigma (Darmstadt, Germany). The experimental water was ultrapure water with a resistivity of 18.2 MΩ/cm, and it was purified using a Millipore ultrapure water system (Burlington, NJ, USA). Additionally, ACS-grade chloroform was sourced from Wokai (Beijing, China). These solutions and reagents provided a solid guarantee for the accurate conduct of the experiment.

2.3.2. Metabolite Extraction

The rats were euthanized using pentobarbital sodium at a dose of 150 mg/kg, followed by the careful removal of their brains. To access the striatum, forceps were applied to the medial cortex. Approximately 100 mg (±2%) of striatal tissue was then transferred into a high-throughput tissue grinder (Meibi, Jiaxing, China) containing 1 mL of tissue extraction buffer (composed of 75% methanol/chloroform 9:1 and 25% H_2O) kept at −20 °C. Three steel balls with diameters of 3 mm were added to facilitate the grinding process. The tissue was ground at 50 Hz for 60 s, which was repeated twice, followed by sonication at room temperature for 30 min. Subsequently, the samples were centrifuged at 12,000 rpm for 10 min at 4 °C, and 850 μL of the resultant supernatant was carefully extracted and concentrated under vacuum until it reached a dry state. For further analysis by liquid chromatography-mass spectrometry (LC-MS), 200 μL of a solution consisting of 50% acetonitrile in 2-chloro phenylalanine (20 ppm) was added to 200 μL of 50% acetonitrile. The resulting mixture was then filtered through a 0.22 μm membrane.

2.3.3. Chromatographic Conditions

The liquid chromatography (LC)-tandem mass spectrometry (MS/MS) system consists of a Waters UPLC (Dublin, Ireland) separation module coupled to a Thermo LTQ (Waltham, MA, USA) ion trap mass spectrometer. An ACQUITY UPLC® HSS T3 (2.1 × 150 mm, 1.8 μm, Waters, Dublin, Ireland) column was used with a sample temperature of 8 °C at room temperature and 40 °C at a flow rate of 0.25 mL/min. A gradient elution was performed using a mobile phase consisting of 0.1% formic acid in water (A2) and 0.1% formic acid in acetonitrile (B2). The gradient elution procedure consisted of the following steps: ① 0–1 min, 2–50% B2; ② 9.5–14 min, 50–98% B2; ③ 14–15 min, 98% B2; ④ 15–15.5 min, 98–2% B2; and ⑤ 15.5–17 min, 2% B2.

2.3.4. Mass Spectrometric Conditions

An electrospray ionization source (ESI) was employed using both the positive and negative ionization modes. The positive ion spray voltage was set at 4.80 kV, while the negative ion spray voltage was set at 4.50 kV. The sheath gas was maintained at 45 arb, and an auxiliary gas was used at 15 arb. The capillary temperature was set at 325 °C. A full scan was conducted at a resolution of 60,000, with a scan range of 89–1000 for positive ions and 114–1000 for negative ions. High-energy collision-induced dissociation (HCD) was applied for secondary cleavage, utilizing a collision voltage of 30 eV. Additionally, dynamic exclusion was implemented to eliminate unnecessary MS/MS information.

2.3.5. Data Analysis

Peak identification was performed using the XCMS package (version v3.8.2) of the Proteowizard software (version v3.0.8789) and R (version v3.3.2). Additionally, dynamic exclusion was implemented to eliminate unnecessary MS/MS information. This analysis provided essential information, including the mass-to-nucleus ratio (m/z), retention time (rt), and peak area (intensity), which was then normalized using sum peak area normalization. The complete dataset can be accessed at MetaboLights (MTBLS8577).

Metabolite identification was conducted utilizing public spectral databases such as HMDB, MassBank, LipidMaps, mzCloud, and KEGG, as well as the proprietary standard compound library established by Nuomi Metabolomics. The parameter was set with a mass deviation of less than 30 ppm to obtain qualitative metabolite results. The specific principle involved determining the molecular weights of the metabolites based on the m/z of the parent ions in the primary mass spectrum. Molecular formulas were predicted using information on the mass deviation and adduct ions, and then they were matched against the databases to achieve primary metabolite identification. Simultaneously, for the metabolites with detected secondary mass spectra listed in the quantification results, fragment ions and other relevant information were matched with each metabolite in the databases to achieve secondary metabolite identification.

The supervised pattern recognition method was used, and a partial least squares-discriminant analysis (PLS-DA) was performed on each group of data. The parameters of R2X, R2Y, and Q2 were extracted. For data processing, the pheatmap package in R was used for scaling and plotting, applying agglomerative hierarchical clustering with the Euclidean distance as the distance metric. Hierarchical clustering was performed based on the relative values of the metabolites under different experimental conditions, and the results were presented in heat maps. A log10 transformation was applied to the raw data, and clustering for both rows and columns was carried out using the Canberra algorithm. The metabolites were considered significantly different at a p-value < 0.05 and a variable importance in projection score (VIP) of >1. The MetPA database was used to analyze the relevant metabolic pathways for each group of differential metabolites, the hypergeometric test algorithm was used for the data analysis, and relative-betweenness centrality was used for the pathway topology analysis.

3. Results

3.1. LC-MS Analysis of the Striatal Metabolome

In this study, LC-MS metabolomics was employed to analyze the metabolite compositions and contents in striatal tissue samples obtained from three distinct groups of rats. These groups included *Shank3*$^{-/-}$ rats that underwent early-life swimming exercise intervention, as well as *Shank3*$^{-/-}$ and *Shank3*$^{+/+}$ control rats that did not receive the intervention.

The BPC trend was consistent for all QC samples, indicating perfect reproducibility and reliable data (Figure 2). Remarkably, the retention times and intensity profiles of the sample peaks exhibited excellent reproducibility, indicating both the high quality of the samples and the reliability of our analytical methodology. In addition, a visual inspection of the chromatograms revealed discernible differences among the various experimental groups, warranting further in-depth analysis and investigation.

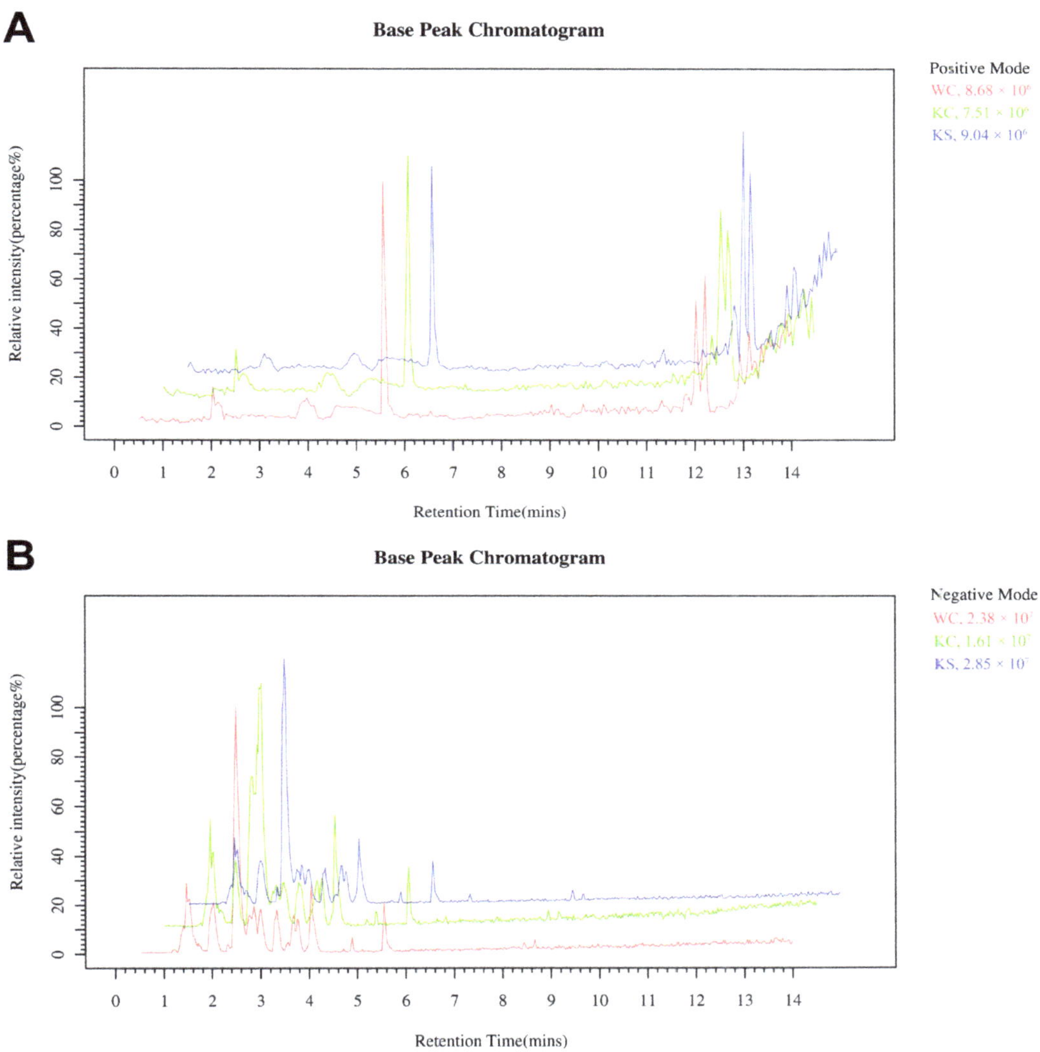

Figure 2. Ion-mode base peak chromatogram. (**A**) A chromatogram of the positive ion mode base peak, and (**B**) a chromatogram of the negative ion mode base peak.

3.2. PLS-DA

The effectiveness of the current PLS-DA model was assessed using permutation graphs to determine the risk of overfitting. The results, as illustrated in Figure 3, indicated significant differences in the combined positive and negative ions as well as the metabolite profiles among the WC, KC, and KS groups. Specifically, the combined positive and negative ion PLS-DA model exhibited R^2Y (cum) = 1 and Q^2 (cum) = 0.66. These values indicated a highly robust and reliable fit, with R^2Y (cum) = 1 suggesting a perfect explanation of the variance in the response variable by the model and Q^2 (cum) = 0.66 indicating a good predictive power and model validity.

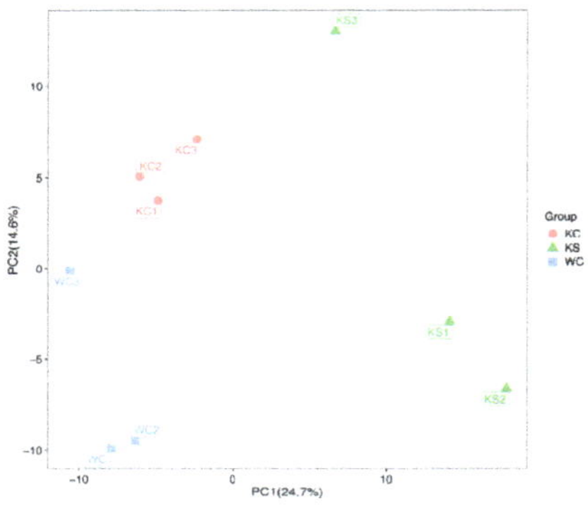

Figure 3. Plots of the PLS-DA scores, loadings, and replacement tests for each comparison group in the positive and negative ion modes.

3.3. *Differential Metabolite Screening*

Differential metabolite screening was conducted using stringent criteria, specifically, a p-value < 0.05 and a VIP of >1. These criteria were applied to both the positive and negative ion pattern metabolites in the following three rat striatal metabolism groups: WC, KC, and KS. Subsequently, heat maps representing the differential metabolites of the ions were generated (Figure 4). Following an annotation analysis of the ion pattern differential compounds, a comprehensive list of the differential metabolites was compiled. In the KC and WC groups, a total of 17 differential metabolites were identified, with 14 metabolites showing down-regulation. In the KS and KC groups, 19 differential metabolites were detected, with 18 metabolites exhibiting up-regulation. Similarly, in the KS and WC groups, 22 differential metabolites were found, with 18 metabolites displaying up-regulation. A summary of these findings is presented in Table 1.

Table 1. Differential metabolite profiles of the striatal tissue samples in the WC, KC, and KS groups.

No.	Metabolites	*m/z*	RT (s)	Formula	KEGG	Type	Fold Change		
							KC/WC	KS/KC	KS/WC
1	Taurine	124.01	89.03	$C_2H_7NO_3S$	C00245	[M−H]−	0.92 *	1.23 *	1.13 *
2	Pyroglutamic acid	128.04	177.84	$C_5H_7NO_3$	C01879	[M−H]−	0.88	1.57 *	1.39 *
3	(E)-Glutaconate	129.02	103.77	$C_5H_6O_4$	C02214	[M−H]−	0.31 *	4.53 *	1.43 *
4	Creatine	130.06	99.11	$C_4H_9N_3O_2$	C00300	[M−H]−	0.88 *	1.29 *	1.14 *
5	L-Malic acid	133.01	119.43	$C_4H_6O_5$	C00149	[M−H]−	0.84 *	1.62 *	1.36 *
6	Threonic acid	135.03	97.76	$C_4H_8O_5$	C01620	[M−H]−	0.74 *	2.15 *	1.59 *
7	L-Glutamine	145.06	88.52	$C_5H_{10}N_2O_3$	C00064	[M−H]−	0.86 *	1.58 *	1.36 *
8	L-Glutamic acid	146.05	92.43	$C_5H_9NO_4$	C00025	[M−H]−	0.86 *	1.53 *	1.32 *
9	3-Hydroxymethylglutaric acid	161.05	239.48	$C_6H_{10}O_5$	C03761	[M−H]−	0.78 *	2.66 *	2.08 *
10	Myo-Inositol	161.05	556.35	$C_6H_{12}O_6$	C00137	[M−H2O−H]−	0.61 *	1.25	0.76 *
11	L-Phenylalanine	166.09	280.18	$C_9H_{11}NO_2$	C00079	[M+H]+	1.06	2.16 *	2.29 *
12	(2R)-2-Hydroxy-3-propanoate	166.98	105.63	$C_3H_7O_7P$	C00197	[M−H2O−H]−	1.05	2.04 *	2.14 *
13	N-Acetyl-L-aspartic acid	174.04	151.21	$C_6H_9NO_5$	C01042	[M−H]−	1.41 *	1.95 *	2.76 *
14	Alpha-D-Glucose	179.06	763.79	$C_6H_{12}O_6$	C00267	[M−H]−	0.46 *	1.31	0.61 *
15	L-Iditol	181.07	665.47	$C_6H_{14}O_6$	C01507	[M−H]−	0.76 *	1.21	0.92
16	(S)-beta-Tyrosine	181.07	808.76	$C_9H_{11}NO_3$	C21308	[M]−	0.29 *	1.20	0.35 *
17	N-Acetylglutamic acid	188.06	200.43	$C_7H_{11}NO_5$	C00624	[M−H]−	0.84	2.28 *	1.92 *
18	Citric acid	191.02	167.31	$C_6H_8O_7$	C00158	[M−H]−	0.99	1.85 *	1.82 *
19	Sebacic acid	201.11	565.53	$C_{10}H_{18}O_4$	C08277	[M−H]−	0.96	7.66 *	7.32 *
20	Pantothenic acid	218.10	294.11	$C_9H_{17}NO_5$	C00864	[M−H]−	0.59 *	1.64 *	0.97
21	Gamma-Glutamylcysteine	248.96	81.79	$C_8H_{14}N_2O_5S$	C00669	[M−H]−	0.84 *	1.18	0.98
22	N2-gamma-Glutamylglutamine	274.10	101.94	$C_{10}H_{17}N_3O_6$	C05283	[M−H]−	1.10	1.77 *	1.95 *
23	Dehydroepiandrosterone	288.29	724.49	$C_{19}H_{28}O_2$	C01227	[M]+	1.89 *	0.51 *	0.96
24	Oxidized glutathione	611.14	219.81	$C_{20}H_{32}N_6O_{12}S_2$	C00127	[M−H]−	2.20 *	2.05 *	4.52 *
25	Theophylline	179.06	643.65	$C_7H_8N_4O_2$	C07130	[M−H]−	0.88	0.85	0.74 *
26	Hypoxanthine	135.03	170.48	$C_5H_4N_4O$	C00262	[M−H]−	1.24	1.29	1.60 *

"*" indicates a significant difference between the two groups ($p < 0.05$).

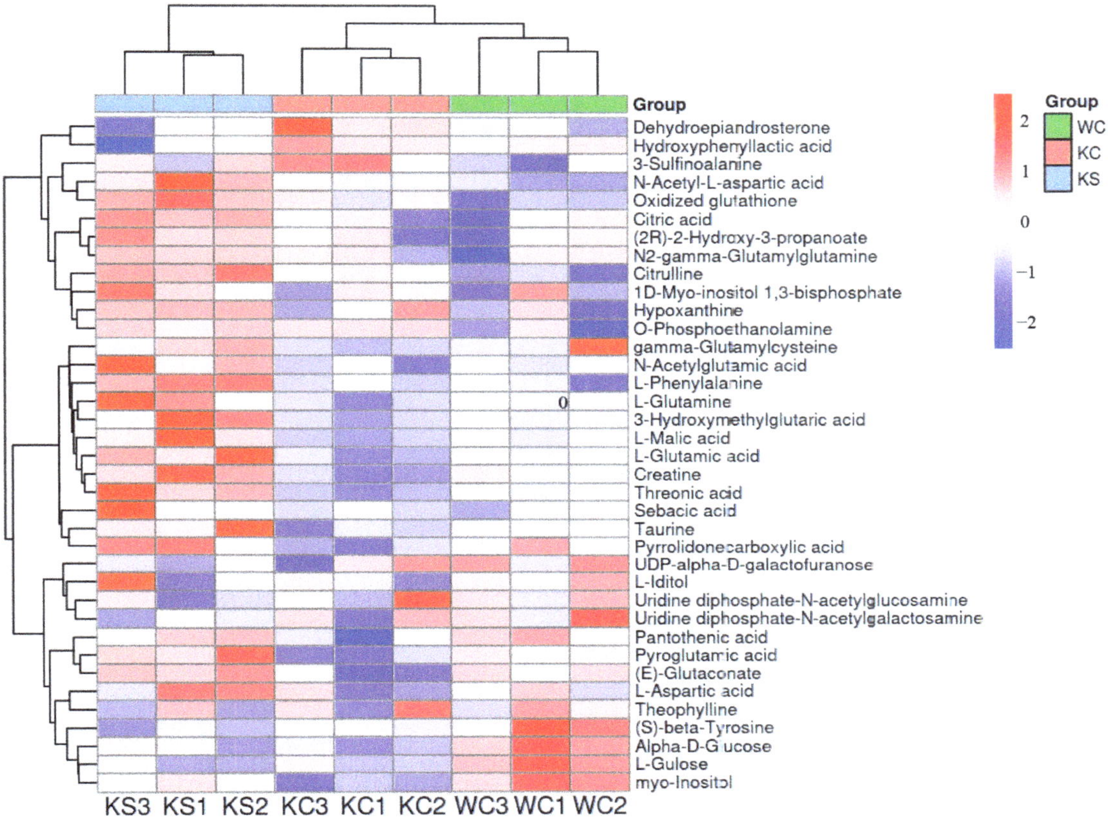

Figure 4. Heat map of the ionic pattern differences in the metabolites.

3.4. Metabolic Network Analysis of the Differential Metabolites

The metabolic pathway analysis of the differential metabolite enrichment is depicted in Figure 5. In the KC and WC groups, the enrichment of the differential metabolites was observed in several key pathways, including the glutamatergic synapse, the GABAergic synapse, alanine, aspartate and glutamate metabolism, D-glutamine and D-glutamate metabolism, ferroptosis, proximal tubule bicarbonate reclamation, taurine and hypo taurine metabolism, nitrogen metabolism, arginine biosynthesis, and glutathione metabolism. Comparatively, the KS vs. KC and KS vs. WC groups exhibited similar patterns of metabolite enrichment pathways, as observed in the KC vs. WC group. However, the KS vs. KC group displayed an additional increase in the central carbon metabolism in cancer pathway, while the KS vs. WC group exhibited an increase in the glucagon-signaling pathway.

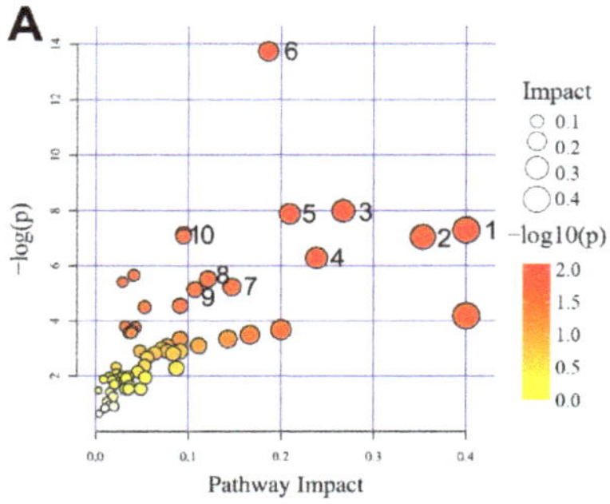

KC vs WC

NO.	Pathway name	Total	Hits	−log(p)	Impact	pathway
1	Glutamatergic synapse	8	2	7.29	0.40	rno04724
2	GABAergic synapse	9	2	7.04	0.35	rno04727
3	Alanine, aspartate and glutamate metabolism	28	3	7.99	0.27	rno00250
4	D-Glutamine and D-glutamate metabolism	13	2	6.28	0.24	rno00471
5	Ferroptosis	29	3	7.88	0.21	rno04216
6	Proximal tubule bicarbonate reclamation	17	4	13.74	0.19	rno04964
7	Taurine and hypotaurine metabolism	22	2	5.22	0.15	rno00430
8	Nitrogen metabolism	19	2	5.52	0.12	rno00910
9	Arginine biosynthesis	23	2	5.14	0.11	rno00220
10	Glutathione metabolism	38	3	7.07	0.09	rno00480

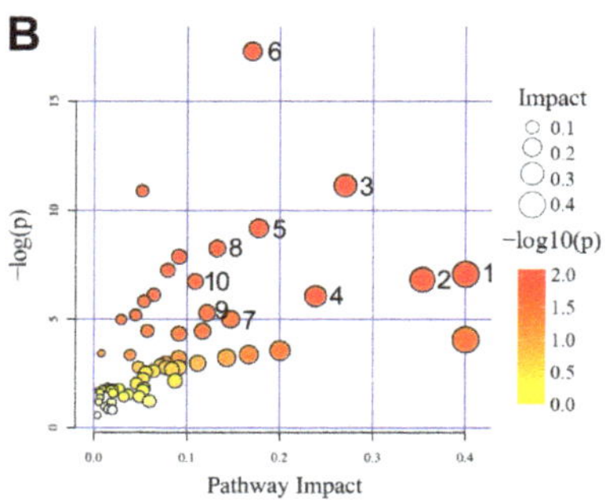

KS vs KC

NO.	Pathway name	Total	Hits	−log(p)	Impact	pathway
1	Glutamatergic synapse	8	2	7.07	0.40	rno04724
2	GABAergic synapse	9	2	6.82	0.35	rno04727
3	Alanine, aspartate and glutamate metabolism	28	4	11.14	0.27	rno00250
4	D-Glutamine and D-glutamate metabolism	13	2	6.06	0.24	rno00471
5	Proximal tubule bicarbonate reclamation	17	3	9.18	0.18	rno04964
6	Central carbon metabolism in cancer	37	6	17.29	0.17	rno05230
7	Taurine and hypotaurine metabolism	22	2	5.00	0.15	rno00430
8	Arginine biosynthesis	23	3	8.24	0.13	rno00220
9	Nitrogen metabolism	19	2	5.29	0.12	rno00910
10	Glutathione metabolism	38	3	6.74	0.11	rno00480

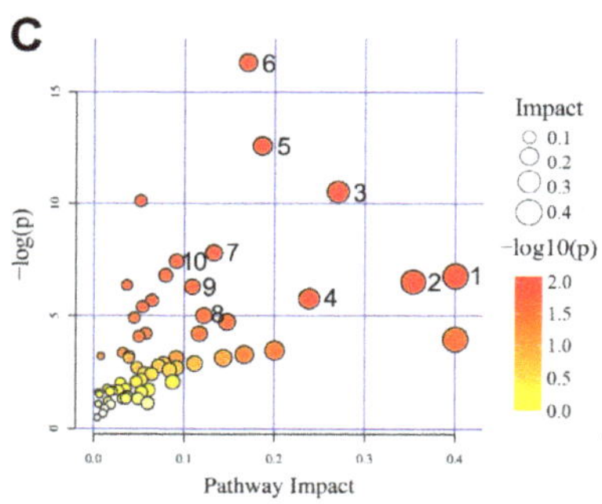

KS vs WC

NO.	Pathway name	Total	Hits	−log(p)	Impact	pathway
1	Glutamatergic synapse	8	2	6.77	0.4	rno04724
2	GABAergic synapse	9	2	6.52	0.3529	rno04727
3	Alanine, aspartate and glutamate metabolism	28	4	10.53	0.2697	rno00250
4	D-Glutamine and D-glutamate metabolism	13	2	5.76	0.2381	rno00471
5	Proximal tubule bicarbonate reclamation	17	4	12.63	0.1858	rno04964
6	Central carbon metabolism in cancer	37	6	16.30	0.1698	rno05230
7	Arginine biosynthesis	23	3	7.79	0.13	rno00220
8	Nitrogen metabolism	19	2	5.00	0.12	rno00910
9	Glutathione metabolism	38	3	6.30	0.1084	rno00480
10	Glucagon signaling pathway	26	3	7.42	0.0909	rno04922

Figure 5. Differential metabolite enrichment pathways in the striatum samples of each comparison group. (**A**) A bubble diagram of the differential metabolite enrichment pathway in the KC vs. WC groups; (**B**) a bubble diagram of the differential metabolite enrichment pathway in the KS vs. KC groups; and (**C**) a bubble diagram of the differential metabolite enrichment pathway in the KS vs. WC groups.

4. Discussion

4.1. Metabolomic Analysis of the Striatum in Shank3 Knockout Rats

ASD represents a complex set of neurodevelopmental disorders with etiological origins rooted in a combination of genetic and environmental factors. Metabolic abnormalities have been identified in several monogenic animal models of ASD, and changes in metabolite levels are emerging as crucial indicators of altered pathway activity in the pathogenesis of ASD. For instance, Rett syndrome, a monogenetic form of ASD resulting from inactivation of the X-linked MECP2 gene encoding the transcription factor methyl cytosine phosphoribosyl guanine binding protein, has been studied extensively [22]. A metabolic analysis of brain extracts from Mecp2-deficient mice, conducted using 1H-NMR, has revealed significant alterations in the D-glutamine and glutamate metabolism [23]. Fragile X syndrome (FXS), the most common genetic cause of ASD, is associated with intellectual disability and results from a recessive X-linked mutation in the fragile X messenger ribonucleoprotein 1 (FMR1) gene [24]. Comprehensive metabolomic studies on the Fmr1-KO mouse model of FXS have demonstrated region-specific metabolic profiles, with the cerebellum and cerebral cortex being particularly affected [25]. These studies have highlighted disruptions in various neurotransmitters, including GABA, glutamate, taurine, and acetylcholine. A metabolite set enrichment analysis (MSEA) revealed significant impacts on the metabolism of D-glutamine, D-glutamate, arginine, and proline [26]. Elevated levels of lipid oxides detected in the cortex of Fmr1-KO mice have indicated increased oxidative stress [26], consistent with previous observations [27,28]. Notably, this metabolic profile exhibited a striking overlap with the pattern detected in the Rett mouse model [23].

In this study, we conducted the first-ever metabolomic analysis of striatal brain regions in a *Shank3* knockout animal model. A comparison between the KC group and the WC group revealed that absence of *Shank3* led to alterations in 17 differential metabolites. Among these, 14 metabolites exhibited down-regulation while 3 metabolites showed up-regulation. Consistent with previous findings for the central differential metabolites observed in animal models of Rett and FXS syndromes, the striatum of the *Shank3* knockout rat model displayed disruptions in key metabolites, including glutamate, glutamine, and oxidized glutathione. These alterations suggest disturbances in synaptic signaling and an increase in oxidative stress. Furthermore, our analysis identified disruptions in metabolites such as taurine, threonine, tyrosine, and glucose, indicating potential abnormalities in neurodevelopmental processes. A functional enrichment analysis of these differential metabolites, combined with transcriptomics [10], revealed functional abnormalities associated with neurodevelopmental processes and oxidative stress. Notably, the disruptions in the GABAergic and glutamatergic synaptic function hint at an imbalance in excitatory-inhibitory (E-I) signaling, a factor closely linked to the pathogenesis of ASD [29].

Our analysis further revealed that several key metabolic pathways were enriched in the striatum of the *Shank3* knockout rats compared to the wild-type controls. These pathways, namely, the glutamatergic synapse and ferroptosis, hold significant biological relevance to the physiological conditions of ASD. The glutamatergic synapse pathway involves various metabolites crucial for glutamate synthesis, storage, release, and uptake [30] Disruptions in glutamatergic signaling have been widely reported in ASD, leading to imbalances in E-I balance and synaptic transmission [31]. Our finding that the glutamatergic synapse pathway was enriched in the *Shank3* knockout rats aligned with previous studies showing alterations in glutamate-related metabolites in ASD models. SHANK3 proteins are essential for the formation and stabilization of synapses, particularly glutamatergic synapses [32]. Therefore, the absence of *Shank3* likely disrupts glutamate signaling, contributing to the ASD phenotype. The ferroptosis pathway, a form of cell death driven by iron-dependent lipid peroxidation, involves multiple metabolites that regulate cellular redox balance and lipid homeostasis [33]. Recent studies have implicated ferroptosis in the pathogenesis of several neurodevelopmental disorders, including ASD [34]. Oxidative stress, a hallmark of ASD, can trigger ferroptosis by overwhelming cellular antioxidant defenses [35]. Our results indicate that the ferroptosis pathway is enriched in *Shank3* knockout rats, suggesting that oxidative stress-induced cell death may play a role in the striatal abnormalities observed in this ASD model. The disruption of *Shank3*-mediated synaptic functions may lead to increased oxidative stress, promoting ferroptosis and contributing to neuronal damage and dysfunction.

4.2. Metabolomic Changes in the Striatum of Rats After Early Swimming Interventions

Previous research has explored the potential therapeutic benefits of swimming interventions in animal models of neurological disorders. These studies have unveiled possible mechanisms through which swimming can mitigate neurological disorders, including up-regulation of neurotrophic factors within the central nervous system [36–38], alterations in synaptic structural proteins [39], changes in synaptic receptor expression [37,40], modifications in neurotransmitter release [41], and adjustments in multiple pathways that affect synaptic plasticity [39,42,43].

In our earlier work, our team conducted transcriptomic studies demonstrating that early swimming interventions not only upregulated genes associated with neural development, synaptic transmission, and synaptic morphology but also modulated various signaling pathways [10]. To delve deeper into the changes in striatal neurotransmission in *Shank3* knockout rats following early swimming interventions and to understand the mechanisms behind the positive behavioral effects observed in animal models of ASD, we extended our previous findings by investigating alterations in striatal metabolomics after swimming interventions in *Shank3* knockout rats. After early swimming intervention, 19 metabolites were changed in the KS group compared with the KC group, of which 18 metabolites were up-regulated and only 1 metabolite was down-regulated. Crucial metabolites related to neurodevelopment, such as taurine and threonine, as well as signaling molecules like glutamate and glutamine, which were down-regulated in the striatum samples due to the *Shank3* knockout, were found to be restored following early swimming interventions. This indicated that early swimming interventions could mitigate aspects of abnormal neurodevelopment and synaptic signaling resulting from *Shank3* gene dysfunction. A subsequent pathway enrichment analysis of these differential metabolites revealed up-regulation in the metabolites associated with taurine and subtaurine metabolism, alanine and aspartate metabolism, arginine biosynthesis, D-glutamine and D-glutamate metabolism, and glutathione metabolism, all of which are linked to neurodevelopment and synaptic transmission, particularly in glutamatergic and GABAergic synapses. The modulatory effects of the early swimming interventions on neurodevelopment and E-I balance were further corroborated. Additionally, a combined analysis of the metabolomics and transcriptomics in the KS and KC groups unveiled functions related to energy metabolism, including glycolysis/gluconeogenesis, and functions related to ASD disease development, such as purine metabolism.

Furthermore, an analysis of the differential metabolites between the KS and WC groups was conducted, revealing 22 differential metabolites, with 18 exhibiting up-regulation. Of particular note, metabolites such as taurine, creatine, glutamine, and glutamate, which had been down-regulated in the striatum samples from the *Shank3* knockout rats, were found to be restored following early swimming interventions. These findings strongly suggest that abnormalities in related metabolites caused by *Shank3* knockout can be effectively reversed by early swimming interventions, potentially leading to improvements in neural development and signaling processes.

The differences in the metabolic enrichment pathways observed between the KS vs. KC and KS vs. WC groups provided deeper insights into the mechanisms by which early swimming interventions may alleviate ASD-like symptoms in *Shank3* knockout rats. In the KS vs. KC group comparison, the primary enrichment pathways included those related to synaptic function (e.g., glutamatergic synapse and GABAergic synapse), energy metabolism (e.g., central carbon metabolism in cancer), and antioxidant pathways (e.g., glutathione metabolism). The up-regulation of metabolites involved in the glutamatergic and GABAergic synapses suggests that early swimming interventions help restore the balance of excitatory and inhibitory neurotransmission, which is vital for normal brain function [44]. Moreover, the enrichment in the central carbon metabolism in the cancer pathway may have indicated enhanced energy metabolism in the striatum samples from the rats undergoing swimming exercise, potentially supporting neuronal recovery and synaptic plasticity [45]. The increase in the glutathione metabolism pathway could reflect a reduction in oxidative stress, a recognized factor in ASD pathogenesis [46]. In the KS vs. WC group comparison, a similar pattern of metabolic enrichment was observed, with additional emphasis on pathways such as glucagon signaling. The restoration of metabolites related to synaptic function and energy metabolism, as well as the reduction in oxidative stress, aligned with the improvements seen in the KS vs. KC comparison. The enrichment in the glucagon signaling pathway, which plays a role in glucose homeostasis and energy metabolism, may indicate that early swimming interventions not only improve synaptic function but also optimize overall metabolic health in the striatum [47]. These differences in the metabolic enrichment pathways have profound implications for the study's outcomes. The restoration of synaptic function and energy metabolism pathways in the KS group, relative to both the KC and WC groups, suggests that early swimming interventions exert a therapeutic effect on the striatal metabolism of *Shank3* knockout rats. The up-regulation of antioxidants and the reduction in oxidative stress further support the neuroprotective role of swimming exercise in this ASD model.

Collectively, these findings indicate that early swimming interventions can effectively reverse some of the metabolic abnormalities caused by *Shank3* knockout, potentially leading to improvements in synaptic function, neuronal health, and, ultimately, behavioral outcomes in *Shank3* knockout rats. This not only offers a mechanistic explanation for the beneficial effects of swimming interventions in ASD but also identifies the key metabolic pathways that may serve as targets for future therapeutic interventions. Furthermore, the differences in metabolic enrichment between the KS vs. KC and KS vs. WC groups underscore the importance of utilizing both knockout and wild-type control groups in metabolomic studies. While comparisons with knockout controls are essential for isolating the effects of an intervention, comparisons with wild-type controls provide additional insights into the extent to which the intervention can restore normal metabolic function. Overall, the comparative analysis of the metabolic enrichment pathways in this study revealed that early swimming interventions have profound impacts on striatal metabolism in *Shank3* knockout rats, suggesting a potential therapeutic role for swimming exercise in ASD.

5. Conclusions

In summary, the metabolomic findings indicate that *Shank3* knockout results in abnormal striatal neurodevelopment and disrupted glutamatergic and GABAergic synaptic function in rats. Additionally, concerning neurotransmitter release, there is a potential disruption in the excitatory–inhibitory (E-I) balance of synaptic signaling. Early swimming interventions effectively reversed the down-regulation of some essential metabolites associated with neurodevelopment and synaptic signaling, which had been negatively impacted by *Shank3* knockout. This reversal could be a key factor contributing to the improvement in the behavior of *Shank3* knockout rats following early swimming interventions.

Author Contributions: Conceptualization, Z.Z.; formal analysis, Y.M.; data curation, Y.X.; writing—original draft preparation, Y.M.; writing—review and editing, Y.H.; funding acquisition, Z.Z. All authors have read and agreed to the published version of the manuscript.

Funding: This work was supported by the Beijing Natural Science Foundation (7232239) and the Humanities and Social Sciences Project of the Ministry of Education (23YJC890027).

Institutional Review Board Statement: The animal study procedure was agreed to by the Animal Care and Use Committee of Peking University (approval no. LA2021552, approved in October 2021).

Informed Consent Statement: Not applicable.

Data Availability Statement: The complete dataset can be accessed at MetaboLights (MTBLS8577).

Conflicts of Interest: The authors declare no conflicts of interest.

References

1. Knopf, A. Autism prevalence increases from 1 in 60 to 1 in 54: CDC. *Brown Univ. Child Adolesc. Behav. Lett.* **2020**, *36*, 4. [CrossRef]
2. Edition, F. Diagnostic and statistical manual of mental disorders. *Am. Psychiatr. Assoc.* **2013**, *21*, 591–643.
3. Zhao, M.; Chen, S. The Effects of Structured Physical Activity Program on Social Interaction and Communication for Children with Autism. *BioMed Res. Int.* **2018**, *2018*, 1825046. [CrossRef] [PubMed]
4. Jiang, X.; Chen, X.; Su, J.; Liu, N. Prevalence of autism spectrum disorder in mainland china over the past 6 years: A systematic review and meta-analysis. *BMC Psychiatry* **2024**, *24*, 404. [CrossRef]
5. Shahat, A.R.S.; Greco, G. The Economic Costs of Childhood Disability: A Literature Review. *Int. J. Environ. Res. Public Health* **2021**, *18*, 3531. [CrossRef]
6. Qiu, S.; Qiu, Y.; Li, Y.; Cong, X. Genetics of autism spectrum disorder: An umbrella review of systematic reviews and meta-analyses. *Transl. Psychiatry* **2022**, *12*, 249. [CrossRef]
7. Ioannidis, V.; Pandey, R.; Bauer, H.F.; Schön, M.; Bockmann, J.; Boeckers, T.M.; Lutz, A.K. Disrupted extracellular matrix and cell cycle genes in autism-associated Shank3 deficiency are targeted by lithium. *Mol. Psychiatry* **2024**, *29*, 704–717. [CrossRef]
8. Monteiro, P.; Feng, G.P. SHANK proteins: Roles at the synapse and in autism spectrum disorder. *Nat. Rev. Neurosci.* **2017**, *18*, 147–157. [CrossRef]
9. Xu, D.; Meng, Y.; An, S.; Meng, W.; Li, H.; Zhang, W.; Xue, Y.; Lan, X.; Wang, X.; Li, M.; et al. Swimming exercise is a promising early intervention for autism-like behavior in Shank3 deletion rats. *CNS Neurosci. Ther.* **2023**, *29*, 78–90. [CrossRef]
10. Meng, Y.; Xu, D.; Zhang, W.; Meng, W.; Lan, X.; Wang, X.; Li, M.; Zhang, X.; Zhao, Y.; Yang, H.; et al. Effect of Early Swimming on the Behavior and Striatal Transcriptome of the Shank3 Knockout Rat Model of Autism. *Neuropsychiatr. Dis. Treat.* **2022**, *18*, 681–694. [CrossRef]
11. Lee, S.S.; Kim, C.J.; Shin, M.S.; Lim, B.V. Treadmill exercise ameliorates memory impairment through ERK-Akt-CREB-BDNF signaling pathway in cerebral ischemia gerbils. *J. Exerc. Rehabil.* **2020**, *16*, 49–57. [CrossRef] [PubMed]
12. Fragala-Pinkham, M.; Haley, S.M.; O'Neil, M.E. Group aquatic aerobic exercise for children with disabilities. *Dev. Med. Child Neurol.* **2008**, *50*, 822–827. [CrossRef] [PubMed]
13. Pan, C.Y. The efficacy of an aquatic program on physical fitness and aquatic skills in children with and without autism spectrum disorders. *Res. Autism Spectr. Disord.* **2011**, *5*, 657–665. [CrossRef]
14. Ennis, E. The effects of a physical therapy-directed aquatic program on children with autism spectrum disorders. *J. Aquat. Phys. Ther.* **2011**, *19*, 4–10.
15. Kim, H.; Woo, R.S.; Yang, E.J.; Kim, H.B.; Jo, E.H.; Lee, S.; Im, H.; Kim, S.; Kim, H.S. Transcriptomic analysis in the striatum reveals the involvement of Nurr1 in the social behavior of prenatally valproic acid-exposed male mice. *Transl. Psychiatry* **2022**, *12*, 324. [CrossRef]
16. Montanari, M.; Martella, G.; Bonsi, P.; Meringolo, M. Autism Spectrum Disorder: Focus on Glutamatergic Neurotransmission. *Int. J. Mol. Sci.* **2022**, *23*, 3861. [CrossRef]
17. Roberts, B.M.; Lopes, E.F.; Cragg, S.J. Axonal modulation of striatal dopamine release by local γ-aminobutyric acid (GABA) signalling. *Cells* **2021**, *10*, 709. [CrossRef]
18. Nicholson, J.K.; Connelly, J.; Lindon, J.C.; Holmes, E. Metabonomics: A platform for studying drug toxicity and gene function. *Nat. Rev. Drug Discov.* **2002**, *1*, 153–161. [CrossRef]
19. Song, T.J.; Lan, X.Y.; Wei, M.P.; Zhai, F.J.; Boeckers, T.M.; Wang, J.N.; Yuan, S.; Jin, M.Y.; Xie, Y.F.; Dang, W.W.; et al. Altered Behaviors and Impaired Synaptic Function in a Novel Rat Model With a Complete Shank3 Deletion. *Front. Cell. Neurosci.* **2019**, *13*, 111. [CrossRef]
20. de Santana Muniz, G.; da Silva, A.M.A.; Cavalcante, T.C.F.; da Silva França, A.K.; Ferraz, K.M.; do Nascimento, E. Early physical activity minimizes the adverse effects of a low-energy diet on growth and development parameters. *Nutr. Neurosci.* **2013**, *16*, 113–124. [CrossRef]
21. Meng, W.; Xu, D.; Meng, Y.; Zhang, W.; Xue, Y.; Zhen, Z.; Gao, Y. Changes in the urinary proteome in rats with regular swimming exercise. *PeerJ* **2021**, *9*, e12406. [CrossRef] [PubMed]

22. Weng, S.-M.; Bailey, M.E.; Cobb, S.R. Rett syndrome: From bed to bench. *Pediatr. Neonatol.* **2011**, *52*, 309–316. [CrossRef] [PubMed]
23. Viola, A.; Saywell, V.; Villard, L.; Cozzone, P.J.; Lutz, N.W. Metabolic fingerprints of altered brain growth, osmoregulation and neurotransmission in a Rett syndrome model. *PLoS ONE* **2007**, *2*, e157. [CrossRef]
24. Penagarikano, O.; Mulle, J.G.; Warren, S.T. The pathophysiology of fragile x syndrome. *Annu. Rev. Genom. Hum. Genet.* **2007**, *8*, 109–129. [CrossRef]
25. Mientjes, E.; Nieuwenhuizen, I.; Kirkpatrick, L.; Zu, T.; Hoogeveen-Westerveld, M.; Severijnen, L.; Rifé, M.; Willemsen, R.; Nelson, D.; Oostra, B. The generation of a conditional Fmr1 knock out mouse model to study Fmrp function in vivo. *Neurobiol. Dis.* **2006**, *21*, 549–555. [CrossRef]
26. Davidovic, L.; Navratil, V.; Bonaccorso, C.M.; Catania, M.V.; Bardoni, B.; Dumas, M.-E. A metabolomic and systems biology perspective on the brain of the fragile X syndrome mouse model. *Genome Res.* **2011**, *21*, 2190–2202. [CrossRef]
27. Bechara, E.G.; Didiot, M.C.; Melko, M.; Davidovic, L.; Bensaid, M.; Martin, P.; Castets, M.; Pognonec, P.; Khandjian, E.W.; Moine, H. A novel function for fragile X mental retardation protein in translational activation. *PLoS Biol.* **2009**, *7*, e1000016. [CrossRef]
28. de Diego-Otero, Y.; Romero-Zerbo, Y.; Bekay, R.; Decara, J.; Sanchez, L.; Fonseca, F.R.-d.; Arco-Herrera, I d. α-tocopherol protects against oxidative stress in the fragile X knockout mouse: An experimental therapeutic approach for the Fmr1 deficiency. *Neuropsychopharmacology* **2009**, *34*, 1011–1026. [CrossRef]
29. Dickinson, A.; Bruyns-Haylett, M.; Jones, M.; Milne, E. Increased peak gamma frequency in individuals with higher levels of autistic traits. *Eur. J. Neurosci.* **2015**, *41*, 1095–1101. [CrossRef]
30. Mahmoud, S.; Gharagozloo, M.; Simard, C.; Gris, D. Astrocytes Maintain Glutamate Homeostasis in the CNS by Controlling the Balance between Glutamate Uptake and Release. *Cells* **2019**, *8*, 184. [CrossRef]
31. Nisar, S.; Bhat, A.A.; Masoodi, T.; Hashem, S.; Akhtar, S.; Ali, T.A.; Amjad, S.; Chawla, S.; Bagga, P.; Frenneaux, M.P.; et al. Genetics of glutamate and its receptors in autism spectrum disorder. *Mol. Psychiatry* **2022**, *27*, 2380–2392. [CrossRef] [PubMed]
32. Kim, Y.; Ko, T.H.; Jin, C.; Zhang, Y.; Kang, H.R.; Ma, R.; Li, H.; Choi, J.I.; Han, K. The emerging roles of Shank3 in cardiac function and dysfunction. *Front. Cell Dev. Biol.* **2023**, *11*, 1191369. [CrossRef] [PubMed]
33. Tang, D.; Chen, X.; Kang, R.; Kroemer, G. Ferroptosis: Molecular mechanisms and health implications. *Cell Res.* **2021**, *31*, 107–125. [CrossRef]
34. Jiang, P.; Zhou, L.; Zhao, L.; Fei, X.; Wang, Z.; Liu, T.; Tang, Y.; Li, D.; Gong, H.; Luo, Y.; et al. Puerarin attenuates valproate-induced features of ASD in male mice via regulating Slc7a11-dependent ferroptosis. *Neuropsychopharmacology* **2024**, *49*, 497–507. [CrossRef]
35. Liu, X.; Lin, J.; Zhang, H.; Khan, N.U.; Zhang, J.; Tang, X.; Cao, X.; Shen, L. Oxidative Stress in Autism Spectrum Disorder-Current Progress of Mechanisms and Biomarkers. *Front. Psychiatry* **2022**, *13*, 813304. [CrossRef]
36. Sigwalt, A.R.; Budde, H.; Helmich, I.; Glaser, V.; Ghisoni, K.; Lanza, S.; Cadore, E.L.; Lhullier, F.L.R.; de Bem, A.F.; Hohl, A.; et al. Molecular aspects involved in swimming exercise training reducing anhedonia in a rat model of depression. *Neuroscience* **2011**, *192*, 661–674. [CrossRef]
37. Boracı, H.; Kirazlı, Ö.; Gülhan, R.; Yıldız Sercan, D.; Şehirli, Ü.S. Neuroprotective effect of regular swimming exercise on calretinin-positive striatal neurons of Parkinsonian rats. *Anat. Sci. Int.* **2020**, *95*, 429–439. [CrossRef]
38. Gyorkos, A.M.; McCullough, M.J.; Spitsbergen, J.M. Glial cell line-derived neurotrophic factor (GDNF) expression and NMJ plasticity in skeletal muscle following endurance exercise. *Neuroscience* **2014**, *257*, 111–118. [CrossRef]
39. Liu, W.; Xue, X.; Xia, J.; Liu, J.; Qi, Z. Swimming exercise reverses CUMS-induced changes in depression-like behaviors and hippocampal plasticity-related proteins. *J. Affect. Disord.* **2018**, *227*, 126–135. [CrossRef]
40. Ko, I.-G.; Kim, S.-E.; Kim, T.-W.; Ji, E.-S.; Shin, M.-S.; Kim, C.-J.; Hong, M.-H.; Bahn, G.H. Swimming exercise alleviates the symptoms of attention-deficit hyperactivity disorder in spontaneous hypertensive rats. *Mol. Med. Rep.* **2013**, *8*, 393–400. [CrossRef]
41. Leite, H.R.; Mourão, F.A.G.; Drumond, L.E.; Ferreira-Vieira, T.H.; Bernardes, D.; Silva, J.F.; Lemos, V.S.; Moraes, M.F.D.; Pereira, G.S.; Carvalho-Tavares, J.; et al. Swim training attenuates oxidative damage and promotes neuroprotection in cerebral cortical slices submitted to oxygen glucose deprivation. *J. Neurochem.* **2012**, *123*, 317–324. [CrossRef] [PubMed]
42. Aguiar Jr, A.S.; Castro, A.A.; Moreira, E.L.; Glaser, V.; Santos, A.R.; Tasca, C.I.; Latini, A.; Prediger, R.D. Short bouts of mild-intensity physical exercise improve spatial learning and memory in aging rats: Involvement of hippocampal plasticity via AKT, CREB and BDNF signaling. *Mech. Ageing Dev.* **2011**, *132*, 560–567. [CrossRef] [PubMed]
43. Anand, S.; Devi, S.A.; Ravikiran, T. Differential expression of the cerebral cortex proteome in physically trained adult rats. *Brain Res. Bull.* **2014**, *104*, 88–91. [CrossRef] [PubMed]
44. Harmon, T.C.; McLean, D.L.; Raman, I.M. Integration of Swimming-Related Synaptic Excitation and Inhibition by olig2(+) Eurydendroid Neurons in Larval Zebrafish Cerebellum. *J. Neurosci.* **2020**, *40*, 3063–3074. [CrossRef]
45. Newman, A.C.; Maddocks, O.D.K. One-carbon metabolism in cancer. *Br. J. Cancer* **2017**, *116*, 1499–1504. [CrossRef]

46. Kwon, D.H.; Cha, H.J.; Lee, H.; Hong, S.H.; Park, C.; Park, S.H.; Kim, G.Y.; Kim, S.; Kim, H.S.; Hwang, H.J.; et al. Protective Effect of Glutathione against Oxidative Stress-induced Cytotoxicity in RAW 264.7 Macrophages through Activating the Nuclear Factor Erythroid 2-Related Factor-2/Heme Oxygenase-1 Pathway. *Antioxidants* **2019**, *8*, 82. [CrossRef]
47. Jiang, G.; Zhang, B.B. Glucagon and regulation of glucose metabolism. *Am. J. Physiol. Endocrinol. Metab.* **2003**, *284*, E671–E678. [CrossRef]

MDPI AG
Grosspeteranlage 5
4052 Basel
Switzerland
Tel.: +41 61 683 77 34

Metabolites Editorial Office
E-mail: metabolites@mdpi.com
www.mdpi.com/journal/metabolites

www.ingramcontent.com/pod-product-compliance
Lightning Source LLC
Chambersburg PA
CBHW062056220926
51850CB00010B/2785
* 9 7 8 3 7 2 5 8 7 7 3 1 7 *